BO LOU
Ferris State University

WILSON
BUFFA
LOU

SIXTH EDITION

STUDENT STUDY GUIDE AND SELECTED SOLUTIONS MANUAL

COLLEGE PHYSICS

VOLUME 2

PEARSON

Prentice
Hall

Upper Saddle River, NJ 07458

Project Managers: Christian Botting, Kathleen Boothby Sestak
Senior Editor: Erik Fahlgren
Editor-in-Chief, Science: Dan Kaveney
Editorial Assistant: Jessica Berta
Executive Managing Editor: Kathleen Schiaparelli
Assistant Managing Editor: Karen Bosch
Production Editor: Ashley M. Booth
Supplement Cover Manager: Paul Gourhan
Supplement Cover Designer: Christopher Kossa
Manufacturing Buyer: Ilene Kahn
Manufacturing Manager: Alexis Heydt-Long

Cover Photography: Greg Epperson / Index Stock Imagery

© 2007 Pearson Education, Inc.
Pearson Prentice Hall
Pearson Education, Inc.
Upper Saddle River, NJ 07458

Printed in the United States of America

10 9 8 7 6 5 4 3 2

ISBN 0-13-174405-4

Pearson Education Ltd., London
Pearson Education Australia Pty. Ltd., Sydney
Pearson Education Singapore, Pte. Ltd.
Pearson Education North Asia Ltd., Hong Kong
Pearson Education Canada, Inc., Toronto
Pearson Educación de Mexico, S.A. de C.V.
Pearson Education—Japan, Tokyo
Pearson Education Malaysia, Pte. Ltd.

Preface

This **Study and Guide and Student Solutions Manual** for "COLLEGE PHYSICS" by Jerry D. Wilson, Anthony J. Buffa, and Bo Lou, sixth edition, was prepared to help students gain a greater understanding of the principles of their introductory physics courses. Most of us learn by summary and examples, and this manual has been organized along these lines. For each chapter you will find:

- **Chapter Objectives**

    states the learning goals for the chapter. The objectives tell you what you should know upon completion of chapter study. Your instructor may omit some topics.

- **Chapter Summary and Discussion**

    outlines the important concepts and provide a brief overview of the major chapter contents. Extra worked out examples and integrated examples are included to further strengthen the concepts and principles. This review allows you to check the thoroughness of your study and serves as a last minute quick review (about 30 minutes) before quizzes or tests. Common students' mistakes and misconceptions are noted.

- **Mathematical Summary**

    lists the important mathematical equations in the chapter. The purpose is for self-review. You should identify each symbol in an equation and explain what relationship the equation describes. The equation number in the text is included for reference. A last glance of this section is helpful before taking a test or quiz.

- **Solutions to Selected Exercises and Paired Exercises**

    provides the worked out solutions of the even-numbered paired (red dot) end-of-chapter text exercises. The paired exercises are similar in nature. You should try to work out the even-numbered paired exercise independently and then check your work with solutions in this manual. After working the odd-numbered exercise of a pair, you can check your answer in the Answers to Odd-Numbered Exercises at the back of the text. Solutions for some additional end-of-the-chapter exercises are also included.

- **Practice Quiz**

    consists of multiple choice questions and exercises of the most fundamental concepts and problem-solving skills in the chapter. This allows you to self-check your understanding and knowledge of the chapter. The answers are also given.

As you can see, this manual provides a through review for each chapter. The conscientious student can make good use of the various sections to assist in understanding and mastering the course contents and preparing for exams. I certainly hope you find this manual helpful in your learning.

ACKNOWLEDGMENT

I would like to thank the following people and organizations for their enormous help and support.

First to the co-authors, Jerry D. Wilson and Anthony J. Buffa, for their numerous constructive and helpful comments and discussions.

To Prentice Hall for the financial support and the editors and staff, Erik Fahlgren, Christian Botting, Kathy Sestak, and Jessica Berta for their guidance and assistance.

To Amanda Phillips for proofreading this manual.

Last but not the least, to my family, Lingfei and Alina, for their essential and generous support and love. I dedicate this manual to them.

Bo Lou, Ph.D., Professor of Physics
Department of Physical Sciences
Ferris State University
Big Rapids, MI 49307
loub@ferris.edu

TABLE OF CONTENTS

CHAPTER 15

Electric Charge, Forces, and Fields

I.   Chapter Objectives

Upon completion of this chapter, you should be able to:

1.   distinguish between the two types of electric charge, state the charge-force law that operates between charged objects, and understand and use the law of charge conservation.

2.   distinguish between conductors and insulators, explain the operation of the electroscope, and distinguish among charging by friction, conduction, induction, and polarization.

3.   understand Coulomb's law and use it to calculate the electric force between charged particles.

4.   understand the definition of the electric field, plot electric field lines, and calculate electric fields for simple charge distributions.

5.   describe the electric field near the surface and in the interior of a conductor, determine where charge accumulates on a charged conductor, and sketch the electric field line outside a charged conductor.

*6.   state the physical basis of Gauss's law, and use the law to make qualitative predictions.

II.   Chapter Summary and Discussion

1.   Electric Charge (Section 15.1)

Electric charge is the property of an object that determines its electrical behavior: the electric force it can exert and the electric force it can experience. There are two types of charges, which are distinguished as positive (+) or negative (−). A positive charge is arbitrarily associated with the proton, and a negative charge with the electron. The magnitude of electron charge (fundamental unit of charge) is $|e| = 1.60 \times 10^{-19}$ C, where the SI unit of charge is the coulomb (C). Both electrons and protons carry this charge but with opposite signs.

The directions of the electric forces on the charges of mutual interaction are given by the **law of charges** or the **charge-force law**: like charges repel and opposite charges attract.

An object with a **net charge** means that it has an *excess* of either positive or negative charges. The law of **conservation of charge** states that the net charge of an isolated system remains constant.

Example 15.1   A piece of glass has a net charge of $-2.00\ \mu C$.

   (a) Are there more protons or electrons in the glass?

   (b) What happens if an identically charged piece of glass is placed near the first one?

   (c) How many fundamental units of excess charge does one piece of the glass contain?

Solution: Given: $q = -2.00\ \mu C = -2.00 \times 10^{-6}\ C$.   Find: (c) n.

   (a) Because the glass has a negative charge, it must have more electrons than protons. Thus, the piece of glass has gained some electrons, since protons are in the atomic nucleus and stay fixed in a solid.

   (b) From the charge-force law, like charges repel. The two pieces of glass therefore repel each other.

   (c) From $q = \pm ne$, we have $n = -\dfrac{q}{|e|} = -\dfrac{-2.00 \times 10^{-6}\ C}{1.60 \times 10^{-19}\ C/electron} = 1.25 \times 10^{13}$ electrons.

2.   Electrostatic Charging (Section 15.2)

A **conductor** is a material that has the ability to conduct or transmit electric charges, whereas an **insulator** is a poor electric conductor.

Electrostatic charging can be accomplished by the following four processes. In **charging by friction**, insulators are rubbed with different materials, and the insulators and the materials acquire equal but opposite charges. In **charging by contact** or **by conduction**, a charged object makes a contact with an uncharged object, and some of the charge on the charged object is transferred to the uncharged object. In **charging by induction**, a charged object is brought near (but not touching) a uncharged object, the uncharged object is then "grounded," and the uncharged object acquires a charge opposite that of the charged object. In **charging by polarization**, the positive and negative charges are simply separated or realigned within the object, and the net charge of the object is still zero. Charges are merely separated, so a portion of the object has excess positive charges and another has excess negative charges.

3.   Electric Force (Section 15.3)

The magnitude of the electric force between two point charges q_1 and q_2 is given by **Coulomb's law**:

$F_e = \dfrac{kq_1q_2}{r^2}$, where $k \approx 9.00 \times 10^9\ N{\cdot}m^2/C^2$ is called the Coulomb constant, and r is the distance between the two point charges. The direction of the electric force is determined from the charge-force law. Like the universal law of gravity $\left(F_g = \dfrac{Gm_1m_2}{r^2}\right)$, Coulomb's law is also an *inverse square law* because the force is proportional to $\dfrac{1}{r^2}$.

Note: Because force is a vector, you have to use vector addition to calculate the net force on a charge if there is more than one force on it (from several other charges).

Example 15.2   Two charges are separated by a distance d and exert mutual attractive forces of F_1 on each other. If the charges are separated by a distance of $3d$, what are the new mutual attractive forces?

Solution: Given: $r_1 = d$, $r_2 = 3d$, and F_1. Find: F_2.

From Coulomb's law, $F_e = \dfrac{kq_1q_2}{r^2}$, we have $\dfrac{F_2}{F_1} = \dfrac{r_1^2}{r_2^2}$ (since k, q_1, and q_2 remain constant).

So $F_2 = \dfrac{r_1^2}{r_2^2} F_1 = \dfrac{(d)^2}{(3d)^2} F_1 = \frac{1}{9}F_1$. The force decreases by a factor of 9 when the distance is increased to 3 times its original value. This is an important feature of the inverse square law.

Example 15.3   A +4.0-C charge is at the origin ($x = 0$), and a +9.0-C charge is at $x = 4.0$ m. Where can a third charge q_3 be placed on the x-axis so the net force on it is zero?

Solution: Given: $q_1 = 4.0$ C, $q_2 = 9.0$ C, $r_1 = d$, $r_2 = (4.0 \text{ m} - d)$, $F_{31} = F_{32}$.

               Find: d.

By the charge-force law, the third charge has to be placed *between* the two charges. The third charge, however, can be either positive or negative. (Why?) Assume the third charge is placed at a distance d from the +4.0 C-charge, then it is $(4.0 \text{ m} - d)$ from the +9.0-C charge. For the force on q_3 to be zero, the forces by q_1 and q_2 on q_3 must equal each other in magnitude. That is, $F_{31} = F_{32}$.

From Coulomb's law, we have $F_{31} = \dfrac{kq_1q_3}{d^2}$ and $F_{32} = \dfrac{kq_2q_3}{(4.0 \text{ m} - d)^2}$.

Equating F_{31} and F_{32} gives $\dfrac{k(4.0 \text{ C})q_3}{d^2} = \dfrac{k(9.0 \text{ C})q_3}{(4.0 \text{ m} - d)^2}$.

Simplifying and taking the square root on both sides, we obtain $\dfrac{2.0}{d} = \dfrac{3.0}{(4.0 \text{ m} - d)}$.

Cross-multiplying, we arrive at $2.0(4.0 \text{ m} - d) = 3.0d$, or $5.0d = 8.0$ m.

Solving for $d = \dfrac{8.0 \text{ m}}{5.0} = 1.6$ m from the +4.0-C charge.

Example 15.4 Consider three point charges located at the corners of a triangle as shown in the accompanying figure. If $q_1 = 6.0$ nC, $q_2 = -1.0$ nC, and $q_3 = 5.0$ nC, what is the net force on q_3.

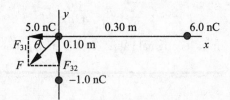

$q_3 = 5.0$ nC   0.30 m $q_1 = 6.0$ nC

0.10 m

$q_2 = -1.0$ nC

Solution: Given: $q_1 = 6.0$ nC $= 6.0 \times 10^{-9}$ C, $q_2 = -1.0$ nC $= -1.0 \times 10^{-9}$ C,

$q_3 = 5.0$ nC $= 5.0 \times 10^{-9}$ C, $r_1 = 0.30$ m, $r_2 = 0.10$ m.

Find: $\vec{\mathbf{F}}$ (net force vector).

By the charge-force law, the force on the 5.0-nC charge by the 6.0-nC charge is to the left, and the force by the -3.0-nC charge is down.

From Coulomb's law, we have

$$F_{31} = \frac{(9.00 \times 10^9 \text{ N·m}^2/\text{C}^2)(6.0 \times 10^{-9} \text{ C})(5.0 \times 10^{-9} \text{ C})}{(0.30 \text{ m})^2} = 3.0 \times 10^{-6} \text{ N};$$

$$F_{32} = \frac{(9.00 \times 10^9 \text{ N·m}^2/\text{C}^2)(1.0 \times 10^{-9} \text{ C})(5.0 \times 10^{-9} \text{ C})}{(0.10 \text{ m})^2} = 4.5 \times 10^{-6} \text{ N}.$$

The negative sign of the negative charge does not have to be included in the preceding calculation because the direction of the force is already determined by the charge-force law.

Thus, $F = \sqrt{F_{31}^2 + F_{32}^2} = \sqrt{(3.0 \times 10^{-6} \text{ N})^2 + (4.5 \times 10^{-6} \text{ N})^2} = 5.4 \times 10^{-6} \text{ N}$,

and $\theta = \tan^{-1} \dfrac{F_{32}}{F_{31}} = \tan^{-1} \dfrac{4.5 \times 10^{-6} \text{ N}}{3.0 \times 10^{-6} \text{ N}} = 56°$ below the $-x$-axis.

4. Electric Field (Section 15.4)

The **electric field** is a *vector field* that describes how nearby charges modify the space around them. It is defined as the electric force per unit positive charge: $\vec{\mathbf{E}} = \dfrac{\vec{\mathbf{F}}_{\text{on } q_+}}{q_+}$, where q_+ is a positive test charge. The direction of the electric field at any point is in the direction of the force experienced by the positive test charge. The SI units of the electric field are N/C. The magnitude of the electric field due to a point charge is $E = \dfrac{kq}{r^2}$.

Note: Because electric field is a vector, you must use vector addition to calculate the net field at a point if there is more than one charge contributing to the fields.

Electric lines of force are the imaginary lines formed by connecting electric field vectors at many points. Their closeness and direction indicate the magnitude and direction of the electric field at any point. In general, the electric lines of force originate from positive charges (or infinity if there is no positive charge) and terminate at negative charges (or infinity if there is no negative charge). The number of lines leaving or entering a charge is proportional to the magnitude of that charge.

Note: Because the direction of electric lines of force indicates the direction of the electric field, no two electric lines of force can cross. (Why?)

The electric field between two closed spaced parallel plates of charge $\pm Q$ is a constant. Its magnitude is given by $E = \dfrac{4\pi k Q}{A}$, where Q is the magnitude of the charge on each plate, and A is the surface area of the plate. The direction of the electric field points from the positive plate to the negative plate.

Example 15.5   A 2.0-C charge is 10 m from a small test charge of 0.10 nC.

(a) What is the electric field at the location of the test charge?

(b) What is the direction of the electric field at the location of the test charge?

(c) What is the magnitude of the force experienced by the test charge?

Solution: Given: $q = 2.0$ C, $q_+ = 0.10$ nC $= 1.0 \times 10^{-10}$ C, $r = 10$ m.

Find: (a) E (b) direction of $\vec{E}$ (c) $F_{on\,q_+}$.

(a) $E = \dfrac{kq}{r^2} = \dfrac{(9.00 \times 10^9 \text{ N·m}^2/\text{C}^2)(2.0 \text{ C})}{(10 \text{ m})^2} = 1.8 \times 10^8$ N/C.

(b) Because the 2.0-C charge is a positive charge, the direction of the electric field at the location of the test charge points radially outward from the +2.0-C charge.

(c) From the definition of electric field, $\vec{E} = \dfrac{\vec{F}_{on\,q_+}}{q_+}$, we have

$F_{on\,q_+} = q_+ E = (1.0 \times 10^{-10}$ C$)(1.8 \times 10^8$ N/C$) = 0.018$ N.

Integrated Example 15.6

A charge of 5.0 μC is placed at the 0-cm mark of a meterstick, and a charge of -4.0 μC is placed at the 50-cm mark. (a) The electric field at the 30-cm mark is pointing (1) toward the 0-cm mark, (b) perpendicular to meterstick, or (3) toward the 50-cm mark. Explain. (b) What is the magnitude of the electric field at the 30-cm mark? (c) At what point on the meterstick is the electric field zero?

(a) Conceptual Reasoning:

A sketch here is very important. At the 30-cm position, the electric field due to the (positive) 5.0-μC charge at the 0-cm mark is to the right; the electric field due to the (negative) –4.0-μC charge at the 50-cm mark is also to the right. Therefore the vector addition of these two fields is also to the right or pointing (3) toward the 50-cm mark.

$q_1 = 5.0 \ \mu C$ E_2

30 cm   50 cm

0 cm E_1 $q_2 = -4.0 \ \mu C$

(b) Quantitative Reasoning and Solution:

Given: $q_1 = 5.0 \ \mu C = 5.0 \times 10^{-6}$ C, $q_2 = -4.0 \ \mu C = 4.0 \times 10^{-6}$ C,

$r_1 = 0.30$ m, $r_2 = (0.50 \text{ m} - 0.30 \text{ m}) = 0.20$ m.

Find: (b) E (c) x (where $E = 0$).

(b) As explained in (a), both electric fields by the two charges at the 30-cm mark are pointing to the right, so the magnitude of the net electric field is simply the addition of the two electric fields by the two charges.

From $E = \dfrac{kq}{r^2}$, we have $E_1 = \dfrac{(9.00 \times 10^9 \text{ N·m}^2/\text{C}^2)(5.0 \times 10^{-6} \text{ C})}{(0.30 \text{ m})^2} = 5.0 \times 10^5$ N/C,

and $E_2 = \dfrac{(9.00 \times 10^9 \text{ N·m}^2/\text{C}^2)(4.0 \times 10^{-6} \text{ C})}{(0.20 \text{ m})^2} = 9.0 \times 10^5$ N/C.

The negative sign of the negative charge does not have to be included here because the direction of the electric field is already determined. (If it were a positive charge, the electric field would point toward the 0-cm mark.). Therefore the net electric field is

$E = E_1 + E_2 = 5.0 \times 10^5$ N/C $+ 9.0 \times 10^5$ N/C $= 1.4 \times 10^6$ N/C.

(c) Because the electric fields by the two charges between the 0- and 50-cm marks are in the same direction, there is no point between the 0-cm and 50-cm marks where the electric field is zero. We have to look outside the 0-and 50-cm marks along the meterstick. It is possible to have zero electric field only at a point to the right of the 50-cm mark along the meterstick. (Why is it impossible to have zero electric field at a point to the left of 0-cm mark along the meterstick?) Assume this point is at a distance x from q_1 at the 0-cm mark. Then, the distance from this point to q_2 at the 50-cm mark is $(x - 0.50 \text{ m})$. The electric fields are in opposite directions.

$E_1 = \dfrac{kq_1}{x^2}$ and $E_2 = \dfrac{kq_2}{(x - 0.50 \text{ m})^2}$.

To produce a zero net field, E_1 must equal E_2, which gives $\dfrac{k(5.0 \times 10^{-6} \text{ C})}{x^2} = \dfrac{k(4.0 \times 10^{-6} \text{ C})}{(x - 0.50 \text{ m})^2}$.

Simplifying and taking the square root on both sides, we have $\dfrac{2.24}{x} = \dfrac{2.0}{x - 0.50 \text{ m}}$.

Cross-multiplying, we arrive at $(2.24)(x - 0.50 \text{ m}) = 2.0x$, or $0.24x = 1.12$ m.

Thus, $x = 4.7$ m from the 0-cm mark.

Example 15.7 Four charges occupy the corners of a square as shown in the accompanying figure. Each charge has a magnitude of 3.0 μC, and the side of the square is 0.50 m. Find the electric field at the geometric center of the square.

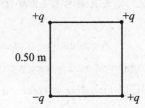

Solution: Given: $q = 3.0 \ \mu C$, $L = 0.50$ m.

Find: $\vec{E}$.

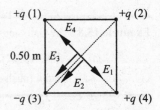

From symmetry, the fields due to the charges at the upper-left (1) and lower-right (4) corners cancel. The fields due to the charges at the upper-right (2) and lower-left corners (3) are in the same direction and pointing toward $-q$ (3), so the net electric field is simply the sum of these two fields.

The length of the diagonal is $d = \sqrt{(0.50 \text{ m})^2 + (0.50 \text{ m})^2} = 0.707$ m.

So the distance from each charge to the center of the square is $\dfrac{d}{2} = \dfrac{0.707 \text{ m}}{2} = 0.354$ m.

From $E = \dfrac{kq}{r^2}$, we have $E_2 = E_3 = \dfrac{(9.00 \times 10^9 \text{ N·m}^2/\text{C}^2)(3.0 \times 10^{-6} \text{ C})}{(0.354 \text{ m})^2} = 2.15 \times 10^{15}$ N/C.

Therefore, $E = E_2 + E_3 = 2(2.15 \times 10^{15} \text{ N/C}) = 4.3 \times 10^{15}$ N/C, toward the negative charge.

5. Conductors and Electric Fields (Section 15.5)

The electric fields associated with charged conductors in electrostatic equilibrium (isolated or insulated) have the following interesting properties:

(1) The electric field is zero everywhere inside a charged conductor.

(2) Any excess charge on an isolated conductor resides entirely on the surface of the conductor.

(3) The electric field at the surface of a charged conductor is perpendicular to the surface.

(4) Excess charge tends to accumulate at sharp points, or locations of the greatest curvature, on charged conductors, so the highest charge accumulations occur where the electric field from the conductor is the largest.

All the preceding properties can be explained with the properties of a conductor and the law of charges.

*6. Gauss's Law for Electric Fields: A Qualitative Approach (Section 15.6)

A qualitative statement of **Gauss's law** is that the net number of electric field lines passing through an imaginary closed surface is proportional to the amount of net charge enclosed within the surface. The surface is called a **Gaussian surface**. In general, by counting the number of lines entering or leaving a Gaussian surface, we can determine the sign and the *relative* magnitude of the charge enclosed within that surface. If there are a net number of lines leaving the surface (we label them "+"), the net charge enclosed is positive. If there are a net number of lines entering the surface (we label them "−"), the net charge enclosed is negative.

Example 15.8   A Gaussian surface encloses an object with a net charge of +2.0 C, and there are 6 lines leaving the surface. Some charge is added to the object, and now there are 18 lines *entering* the surface. How much charge was added?

Solution:

Because there are 6 electric field lines when there is an initial charge of $Q_i = +2.0$ C, a charge of +1.0 C is equivalent to 3 electric field lines. After charge is added, there are 18 electric field lines entering, so the net charge is now

$$Q_f = -\frac{18 \text{ lines}}{3 \text{ lines/coulomb}} = -6.0 \text{ C.}$$

The charge is negative because the electric field lines are entering the surface.

Therefore, the charge added was $\Delta Q = Q_f - Q_i = -6.0 \text{ C} - 2.0 \text{ C} = -8.0 \text{ C.}$

III.   Mathematical Summary

Coulomb's Law (Two Point Charges)	$F_e = \dfrac{kq_1 q_2}{r^2}$ (15.2) $k \approx 9.00 \times 10^9 \text{ N·m}^2/\text{C}^2$	Computes the magnitude of the electric force between two point charges.
Electric Field (Definition)	$\vec{E} = \dfrac{\vec{F}_{\text{on } q_+}}{q_+}$ (15.3)	Defines the electric field vector from the force vector on a positive test charge q_+.
Electric Field due to a Point Charge q	$E = \dfrac{kq}{r^2}$ (15.4)	Calculates the magnitude of the electric field due to a point charge.

IV. Solutions of Selected Exercises and Paired Exercises

8.   Yes, both objects are charged with the same type of charge if they repel each other.

However, if two objects attract each other, both are not necessarily charged. For example, through polarization and polarization by induction, a charged object can attract a neutral object, but the neutral object has a nonuniform charge distribution.

12.   (a) The charge on the silk must be $\boxed{\text{(3) negative}}$ because of the conservation of charge. When one object becomes positively charged, it loses electrons. These same electrons must be gained by another object, and therefore it is negatively charged.

(b) $\boxed{-8.0 \times 10^{-10}\ \text{C}}$ according to charge conservation.

From $q = -ne$, we have $n = -\dfrac{q}{e} = \dfrac{-8.0 \times 10^{-10}\ \text{C}}{1.6 \times 10^{-19}\ \text{C}} = \boxed{5.0 \times 10^{9}\ \text{electrons}}$.

(c) The mass is $m = (9.11 \times 10^{-31}\ \text{kg/electron})(5.0 \times 10^{9}\ \text{electrons}) = \boxed{4.6 \times 10^{-21}\ \text{kg}}$.

18.   No, charges simply reorient themselves. There is no gain or loss of electrons.

26.   (a) The answer is $\boxed{\text{(4) 1/4}}$. Since $F_e = \dfrac{kq_1 q_2}{r^2}$, F is inversely proportional to the square of the distance, r. If r doubles, F_e becomes 1/4 times the original force.

(b) $\dfrac{F_e}{F_{eo}} = \dfrac{r_o^{\ 2}}{r^2} = \dfrac{1^2}{(1/3)^2} = 9$. So $F_e = \boxed{9F}$.

30.   (a) Since $F_e = \dfrac{kq_1 q_2}{r^2}$, $r = \sqrt{\dfrac{kq_1 q_2}{F_e}}$ or r is proportional to $1/\sqrt{F_e}$. If F_e decreases by a factor of 10, r will increase but not by a factor of 10. So the answer is $\boxed{\text{(1) less than 10 times}}$ the original distance.

(b) From $F_e = \dfrac{kq_1 q_2}{r^2}$, we have $\dfrac{F_2}{F_1} = \dfrac{r_1^{\ 2}}{r_2^{\ 2}}$, so $r_2 = \sqrt{\dfrac{F_1}{F_2}}\ r_1 = \sqrt{10}\ (30\ \text{cm}) = \boxed{95\ \text{cm}}$.

33.   (a) By symmetry, the electron has to be at the $\boxed{50\ \text{cm}}$ mark since both forces are repulsive and opposite.

(b) By symmetry, the proton has to be at the $\boxed{50\ \text{cm}}$ mark since both forces are attractive and opposite.

38. $F_2 = F_3 = \dfrac{kq_1q_2}{r^2} = \dfrac{(9.0 \times 10^9 \text{ N·m}^2/\text{C}^2)(4.0 \times 10^{-6} \text{ C})^2}{(0.20 \text{ m})^2} = 3.6 \text{ N}.$

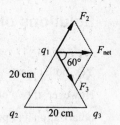

According to symmetry, the net force on q_1 points in the +x direction.

$F_{net} = F_x = F_2 \cos 60° + F_3 \cos 60°$

$= 2(3.6 \text{ N}) \cos 60° = \boxed{3.6 \text{ N in the positive } x \text{ direction}}.$

46. Electric field is defined as the ratio of force to charge at a given point in space. The force on a charge can only point in one direction and so does the electric field. Therefore, the field lines never cross, since that would indicate two different directions at the crossing point.

50. (a) Since $E = \dfrac{kq}{r^2}$, E is inversely proportional to the square of the distance.

So if r doubled, E $\boxed{(2) \text{ decreased}}$.

(b) $\dfrac{E}{E_o} = \dfrac{r_o^2}{r^2} = \dfrac{1^2}{2^2} = 1/4.$ Therefore $E = \dfrac{E_o}{4} = \dfrac{1.0 \times 10^{-4} \text{ N/C}}{4} = \boxed{2.5 \times 10^{-5} \text{ N/C}}.$

52. $E = \dfrac{kq}{r^2} = \dfrac{(9.0 \times 10^9 \text{ N·m}^2/\text{C}^2)(2.0 \times 10^{-12} \text{ C})}{(0.75 \times 10^{-2} \text{ m})^2} = \boxed{3.2 \times 10^2 \text{ N/C away from the charge}}.$

56. (a) Since both charges are negative, a point between the two charges could have zero electric field. The location is at $\boxed{(1) \text{ left of the origin}}$ because the charge closer to the point has the smaller magnitude. The distance from the smaller charge has to be smaller to match the field by the bigger charge.

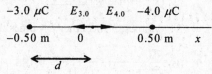

(b) Assume the point is d from the $-3.0\text{-}\mu$C charge. For the electric field to be zero, $E_{3.0} = E_{4.0}$.

So $E_{3.0} = \dfrac{kq}{r^2} = \dfrac{k(3.0 \ \mu\text{C})}{d^2} = E_{4.0} = \dfrac{k(4.0 \ \mu\text{C})}{(1.0 - d)^2}.$

Taking the square root on both sides gives $\dfrac{\sqrt{3.0}}{d} = \dfrac{\sqrt{4.0}}{1.0 - d}$. Or $1.73(1.0 - d) = 2.0d$.

Solving for $d = 0.464$ m. Therefore the coordinates of the point are $\boxed{(-0.036 \text{ m}, 0)}$.

60. Due to symmetry, E_1 and E_2 cancel out. So the net electric field is the electric field by q_3.

$E = E_3 = \dfrac{kq_3}{r^2} = \dfrac{(9.0 \times 10^9 \text{ N·m}^2/\text{C}^2)(4.0 \times 10^{-6} \text{ C})}{[(0.20 \text{ m}) \cos 30°]^2}$

$= \boxed{1.2 \times 10^6 \text{ N/C toward the charge of } -4.0 \ \mu\text{C}}.$

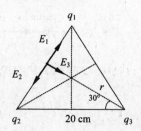

66. (a) From symmetry, the net electric field is in the $\boxed{-y}$ direction.

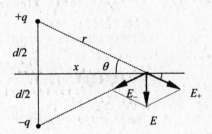

(b) $r = \sqrt{(d/2)^2 + x^2}$, $\sin \theta = \dfrac{d/2}{r} = \dfrac{d/2}{\sqrt{(d/2)^2 + x^2}}$.

$E_- = E_+ = \dfrac{kq}{r^2} = \dfrac{kq}{(d/2)^2 + x^2}$.

$E = E_y = E_- \sin \theta + E_+ \sin \theta = 2E_+ \sin \theta$

$= \dfrac{2kq}{(d/2)^2 + x^2} \dfrac{d/2}{\sqrt{(d/2)^2 + x^2}} = \boxed{\dfrac{kqd}{[(d/2)^2 + x^2]^{3/2}}}$.

(c) If $x \gg d$, we can ignore the term $(d/2)^2$ in $(d/2)^2 + x^2$. So $E = \dfrac{kqd}{[(d/2)^2 + x^2]^{3/2}} \approx \dfrac{kqd}{(x^2)^{3/2}} = \dfrac{kqd}{x^3}$.

(d) This is because the field is not created by a single point charge. We expect inverse square from single point charges, but vector addition can cause other distance dependences as happens here.

70. Yes, because the car (a metal frame) keeps the electric field from reaching you.

73. (a) The inner surface of the shell will have $\boxed{\text{(1) negative}}$ charge due to induction.

(b) $\boxed{\text{Zero}}$ since all excess charge resides on the surface of the conductor in electrostatic equilibrium.

(c) $\boxed{+Q}$ since all excess charge resides on the surface of the conductor in electrostatic equilibrium.

(d) $\boxed{-Q}$ by induction and the conservation of charge.

(e) $\boxed{+Q}$ by induction and the conservation of charge.

84. (a) The magnitude of the unknown charge is $\boxed{\text{(1) greater than 10.0 } \mu C}$ because the number of field lines is proportional to the magnitude of the charge inside the Gaussian surface.

(b) $q_2 = -\dfrac{75}{16}(+10.0 \ \mu C) = \boxed{-46.9 \ \mu C}$.

91. $E_x = \dfrac{kq}{r^2} = \dfrac{(9.00 \times 10^9 \ \text{N·m}^2/\text{C}^2)(4.0 \times 10^{-6} \ \text{C})}{(4.0 \ \text{m})^2} = (2.25 \times 10^3 \ \text{N/C}) \ \hat{\mathbf{x}}$.

$E_y = -\dfrac{(9.00 \times 10^9 \ \text{N·m}^2/\text{C}^2)(5.0 \times 10^{-6} \ \text{C})}{(3.0 \ \text{m})^2} = -(5.00 \times 10^3 \ \text{N/C}) \ \hat{\mathbf{y}}$.

So $E = \sqrt{(2.25 \times 10^3 \ \text{N/C})^2 + (5.00 \times 10^3 \ \text{N/C})^2} = \boxed{5.5 \times 10^3 \ \text{N/C}}$,

and $\theta = \tan^{-1}\left(\dfrac{-5.00}{2.25}\right) = \boxed{66° \text{ below positive } x\text{-axis}}$.

93. (a) $t = \dfrac{x}{v_x} = \dfrac{0.10 \text{ m}}{6.15 \times 10^7 \text{ m/s}} = 1.626 \times 10^{-9}$ s.

From kinematics, $d = \frac{1}{2} a_y t^2$, we have $a_y = \dfrac{2d}{t^2} = \dfrac{2(4.70 \times 10^{-3} \text{ m})}{(1.626 \times 10^{-9} \text{ s})^2} = 3.555 \times 10^{15}$ m/s^2.

From Newton's second law, $a_y = \dfrac{F}{m} = \dfrac{qE}{m}$.

So $E = \dfrac{a_y m}{q} = \dfrac{a_y m}{q} = \dfrac{(3.555 \times 10^{15} \text{ m/s}^2)(9.11 \times 10^{-31} \text{ kg})}{(1.60 \times 10^{-19} \text{ C})} = \boxed{2.02 \times 10^4 \text{ N/C}}$.

(b) Since $E = \dfrac{4\pi k Q}{A}$, the surface charge density is

$\dfrac{Q}{A} = \dfrac{E}{4\pi k} = \dfrac{2.024 \times 10^4 \text{ N/C}}{4\pi (9.00 \times 10^9 \text{ N·m}^2/\text{C}^2)} = \boxed{1.79 \times 10^{-7} \text{ C/m}^2}$.

V. Practice Quiz

1. Two point charges, initially 4.0 cm apart, experience a mutual force of magnitude 1.0 N. If they are moved to a new separation of 1.0 cm, what is the magnitude of the new electric force between them?

    (a) 4.0 N   (b) 16 N   (c) 1/4 N   (d) 1/16 N   (e) none of the above

2. A positively charged rod is brought near one end of an uncharged metal bar. The end of the metal bar farthest from the charged rod will be which of the following?

    (a) positive   (b) negative   (c) neutral   (d) attracted   (e) repelled

3. Two point charges have a value of +3.0 C and each are separated by a distance of 4.0 m. What is the magnitude of the electric field at a point midway between the two charges?

    (a) zero   (b) 9.0×10^7 N/C   (c) 18.0×10^7 N/C   (d) 9.0×10^7 N/C   (e) 4.5×10^7 N/C

4. Two point charges, separated by 1.5 cm, have charge values of +2.0 and −4.0 μC, respectively. What is the magnitude of the electric force between them?

    (a) 160 N   (b) 320 N   (c) 3.6×10^{-8} N   (d) 8.0×10^{-12} N   (e) 3.1×10^{-3} N

5. Which of the following is a vector?

    (a) electric charge   (b) Gaussian surface   (c) electric field   (d) both (a) and (c)   (e) both (b) and (c)

6. Three point charges are located at the following positions: $q_1 = 2.00$ μC at $x = 1.00$ m; $q_2 = 3.00$ μC at $x = 0$; $q_3 = -5.00$ μC at $x = -1.00$ m. What is the magnitude of the force on q_2?

    (a) 5.40×10^{-2} N   (b) 0.135 N   (c) 8.10×10^{-2} N   (d) 0.189 N   (e) 9.0×10^{-2} N

7.   A ball with a charge of $+4.0$ μC has a mass of 5.0×10^{-3} kg. What electric field directed upward will exactly balance the weight of the ball?

(a) 4.1×10^2 N/C   (b) 8.2×10^2 N/C   (c) 1.2×10^4 N/C   (d) 5.1×10^6N/C   (e) 2.0×10^{-7} N/C

8.   If a conductor is in electrostatic equilibrium near an electric charge,

(a) the total charge on the conductor must be zero.

(b) any charge on the conductor must be uniformly distributed.

(c) the force between the conductor and the charge must be zero.

(d) there may be some net excess charge on the surface of the conductor.

(e) the total electric field of the conductor must be zero.

9.   Two point charges of $+3.0$ μC and -7.0 μC are placed at $x = 0$ and $x = 0.20$ m, respectively. What is the magnitude of the electric field at the point midway between them ($x = 0.10$ m)?

(a) 9.0×10^6 N/C   (b) 4.5×10^6 N/C   (c) 3.6×10^6 N/C   (d) 1.8×10^6 N/C   (e) 9.0×10^5 N/C

10.   If the distance between two closely spaced parallel plates is halved, the electric field between them

(a) is also halved.   (b) is reduced to 1/4 as large.   (c) remains the same   (d) is doubled.

(e) becomes 4 times as large.

11.   One Gaussian surface containing a charge of $+1.0$ C has 6 electric field lines leaving the surface. What is the charge in another Gaussian surface if there are 12 electric field lines entering it?

(a) $+0.50$ C   (b) -0.50 C   (c) -1.0 C   (d) $+2.0$ C   (e) -2.0 C

12.   Two point charges of -1.0 C and $+1.0$ C are fixed at opposite ends of a meterstick. Where can you put a third negative charge on the meterstick so that the third charge is in equilibrium?

(a) nowhere   (b) at the 0-cm mark   (c) at the 50-cm mark   (d) at the 100-cm mark   (e) anywhere

Answers to Practice Quiz:

1. b 2. a 3. a 4. b 5. c 6. d 7. c 8. d 9. a 10. c 11. e 12. a

CHAPTER 16

Electric Potential, Energy, and Capacitance

I.   Chapter Objectives

Upon completion of this chapter, you should be able to:

1.   understand the concept of electric potential difference (voltage) and its relationship to electric
      potential energy, and calculate electric potential differences.

2.   explain what is meant by an equipotential surface, sketch equipotential surfaces for simple charge
      configurations, and explain the relationship between equipotential surfaces and electric fields.

3.   define capacitance and explain what it means physically, and calculate the charge, voltage, electric
      field, and energy storage for parallel-plate capacitors.

4.   understand what a dielectric is, and understand how it affects the physical properties of a
      capacitor.

5.   find the equivalent capacitance of capacitors connected in series and in parallel, calculate the
      charge, voltage, and energy storage of individual capacitors in series and parallel configurations,
      and analyze capacitor networks that include both series and parallel arrangements.

II.   Chapter Summary and Discussion

1.   Electric Potential Energy and Electric Potential Difference (Section 16.1)

The **electric potential difference (voltage)** between two points is the work per unit positive charge done
by an external force in moving charge between these two points, or the change in electric potential energy per unit
positive charge: $\Delta V = V_B - V_A = \dfrac{W}{q_+} = \dfrac{\Delta U_e}{q_+}$, where q_o is the positive test charge. The SI units of electric potential
difference are joule/coulomb (J/C), or volt (V).

Note: Potential difference and potential energy difference are *different* quantities. Because potential
difference is defined per unit charge, it does not depend on the amount of charge moved, whereas potential energy
difference does.

In a uniform electric field E, the potential difference in moving a charge through a straight-line distance d
(in a direction opposite the electric field, or from negative plate to positive plate in parallel plates) is $\Delta V = Ed$.

Example 16.1   An electron, initially at rest, is accelerated through an electric potential difference of 50.0 V.

(a) What is the kinetic energy of the electron?

(b) What is the speed of the electron?

Solution: Given: $\Delta V = 50.0$ V, $q = 1.60 \times 10^{-19}$ C, $m = 9.11 \times 10^{-31}$ kg.

Find: (a) K (b) v.

(a) When a negatively charged electron is accelerated through a potential difference, work is done on the electron. The electron gains kinetic energy while losing potential energy. The electron moves from a location of lower potential to one of higher potential but from a location of higher potential energy to one of lower potential energy. Why?

From the conservation of energy, $\Delta K = K - K_o = |\Delta U_e|$, but $K_o = 0$, so $\Delta K = K = |\Delta U_e|$.

Because $\Delta V = \dfrac{\Delta U_e}{q}$, we can calculate the kinetic energy gained by the electron:

$K = |\Delta U_e| = q\Delta V = (1.60 \times 10^{-19}$ C$)(50.0$ V$) = 8.00 \times 10^{-18}$ J.

(b) From $K = \frac{1}{2}mv^2$, we have $v = \sqrt{\dfrac{2K}{m}} = \sqrt{\dfrac{2(8.00 \times 10^{-18}\text{ J})}{9.11 \times 10^{-31}\text{ kg}}} = 4.19 \times 10^6$ m/s.

Example 16.2   A 12-V battery maintains the electric potential difference between two parallel metal plates separated by 0.10 m. What is the electric field between the plates?

Solution: Given: $\Delta V = 12$ V, $d = 0.10$ m.

Find: E.

From $\Delta V = Ed$, we have $E = \dfrac{\Delta V}{d} = \dfrac{12\text{ V}}{0.10\text{ m}} = 1.2 \times 10^2$ V/m.

The electric potential difference due to a point charge is given by $\Delta V = \dfrac{kq}{r_B} - \dfrac{kq}{r_A}$, where r_A and r_B are the distances from the point charge to the two points A and B. If we define the potential at infinity as zero ($V = 0$ at $r = \infty$), then the **electric potential** at a distance r from a point charge is $V = \dfrac{kq}{r}$.

The electric potential changes according to the following general rules:

- *Electric potential increases for locations closer to positive charges or farther from negative charges.*
- *Electric potential decreases for locations farther from positive charges or closer to negative charges.*

Note: Unlike electric field, electric potential is a scalar quantity. When adding potentials due to point charges, you need only to add them algebraically (including the positive and negative signs). Also, electric potential is proportional to $1/r$, while the magnitude of the electric field is proportional to $1/r^2$.

For a configuration of multiple point charges, the total potential energy of the system U_{total} is

$$U_{total} = U_{12} + U_{23} + U_{13} + \ldots, \quad \text{where} \quad U_{ij} = \frac{kq_i q_j}{r_{ij}} \text{ is the potential energy due to any two point charges i and j.}$$

Example 16.3   A charge of 5.0 nC is at (0,0), and a second charge of –2.0 nC is at (3.0 m, 0 m). If the potential is taken to be zero at infinity,

(a) what is the electric potential at point (0, 4.0) m?

(b) what is the potential energy of a 1.0-nC charge at point (0, 4.0) m?

(c) what is the work required to bring a charge of 1.0 nC charge from infinity to point (0, 4.0) m?

(d) what is the total potential energy of the three-charge system?

Solution: Given: $q_1 = 5.0$ nC $= 5.0 \times 10^{-9}$ C, $(x_1, y_1) = (0,0)$;

$q_2 = -2.0$ nC $= -2.0 \times 10^{-9}$ C, $(x_2, y_2) = (3.0,0)$ m;

$q_3 = 1.0$ nC $= 1.0 \times 10^{-9}$ C, $(x_3, y_3) = (0, 4.0)$ m.

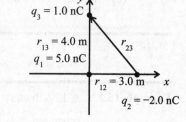

Find: (a) V_{total} (b) U_e (c) W (d) U_{total}.

In the triangle shown, we calculate the distance from q_2 to q_3:

$$r_{23} = \sqrt{(3.0 \text{ m})^2 + (4.0 \text{ m})^2} = 5.0 \text{ m}.$$

(a) From $V = \dfrac{kq}{r}$, the total potential at q_3, V_{total}, is

$$V_{total} = V_1 + V_2 = \frac{(9.00 \times 10^9 \text{ N·m}^2/\text{C}^2)(5.0 \times 10^{-9} \text{ C})}{4.0 \text{ m}} + \frac{(9.00 \times 10^9 \text{ N·m}^2/\text{C}^2)(-2.0 \times 10^{-9} \text{ C})}{5.0 \text{ m}} = 7.65 \text{ V}.$$

(b) From $V = \dfrac{U_e}{q_+}$, we have $U_e = q_3 V = (1.0 \times 10^{-9} \text{ C})(7.65 \text{ V}) = 7.65 \times 10^{-9}$ J.

(c) From conservation of energy, the work required to bring the 1.0-nC charge from infinity is equal to its potential energy, $W = U_e = 7.65 \times 10^{-9}$ J.

(d) $U_{total} = U_{12} + U_{23} + U_{13} = \dfrac{kq_1 q_2}{r_{12}} + \dfrac{kq_2 q_3}{r_{23}} + \dfrac{kq_1 q_3}{r_{13}} = \dfrac{(9.00 \times 10^9 \text{ N·m}^2/\text{C}^2)(5.0 \times 10^{-9} \text{ C})(-2.0 \times 10^{-9} \text{ C})}{3.0 \text{ m}}$

$$+ \frac{(9.00 \times 10^9 \text{ N·m}^2/\text{C}^2)(-2.0 \times 10^{-9} \text{ C})(1.0 \times 10^{-9} \text{ C})}{5.0 \text{ m}}$$

$$+ \frac{(9.00 \times 10^9 \text{ N·m}^2/\text{C}^2)(5.0 \times 10^{-9} \text{ C})(1.0 \times 10^{-9} \text{ C})}{4.0 \text{ m}} = -2.2 \times 10^{-8} \text{ J}.$$

---

2. Equipotential Surfaces and the Electric Field (Section 16.2)

Equipotential surfaces (equipotentials) are surfaces on which a charge experiences a constant electric potential V, so it takes no work to move a charge along an equipotential. The electric field is *perpendicular* to such surfaces at all points. If the equipotentials are given, you can draw the electric field lines by simply connecting arrows perpendicular to the equipotentials; or if the electric field lines are given, you can obtain the equipotentials by drawing surfaces perpendicular to the electric field lines.

The electric field is related to the electric potential difference as follows: $E = \left| \dfrac{\Delta V}{\Delta x} \right|_{max}$. This result means that the electric field is equal to the greatest change in V over the change in x. The units of electric field are volts per meter (V/m), which has the same dimensions as N/C, introduced in Chapter 15.

An **electron volt** (eV) is the kinetic energy gained by an electron accelerated from rest though a potential difference of 1 V. $1\ eV = (1\ e)(1\ V) = (1.60 \times 10^{-19}\ C)(1\ V) = 1.60 \times 10^{-19}\ J$.

Integrated Example 16.4

Two parallel plates, separated by 0.10 m, are connected to a 6.0-V battery. An electron is released from rest at the negative plate and a proton is also released from rest but at the positive plate. (a) When the electron reaches the positive plate and the proton reaches the negative plate, the speed of the proton is (1) higher than, (2) the same as, or (3) lower than the speed than the electron. Explain. (b) What is the electric field between the battery plates? (c) What are the kinetic energies and speeds of the proton and the electron?

(a) Conceptual Reasoning:

The increase in kinetic energy is equal to the change in electric potential energy which depends on the product of charge q and potential difference ΔV ($U_e = q \Delta V$). Since both the electron and proton experience the same potential difference and they both have the same magnitude of charge, they will gain the same amount of kinetic energy going from one plate to the other. However, a proton has more mass than an electron and kinetic energy is $K = \frac{1}{2} mv^2$, the speed of the proton is (3) lower than the speed of the electron.

(b) Quantitative Reasoning and Solution:

Solution: Given: $\Delta V = 6.0\ V$, $\Delta x = 0.10\ m$, $m_e = 9.11 \times 10^{-31}\ kg$, $m_p = 1.67 \times 10^{-27}\ kg$,

$q_e = -1.60 \times 10^{-19}\ C$, $q_p = +1.60 \times 10^{-19}\ C$

Find: (b) E (c) K_e, K_p, and v_e, v_p.

(b) $E = \left| \dfrac{\Delta V}{\Delta x} \right|_{max} = \dfrac{6.0\ V}{0.10\ m} = 60\ V/m$.

(c) From energy conservation, the decrease in the potential energy of the electron is equal to the increase in its kinetic energy, that is, $\Delta K = |\Delta U_e|$. $\Delta K = K - K_o = K - 0 = K = |\Delta U_e| = |q \Delta V|$.

So $K_e = K_p = K = (1.60 \times 10^{-19}\,\text{C})(6.0\,\text{V}) = 9.6 \times 10^{-19}\,\text{J}$ for both electron and proton.

Because $K = \frac{1}{2}mv^2$, $v = \sqrt{\dfrac{2K}{m}}$.

$$v_e = \sqrt{\dfrac{2(9.6 \times 10^{-19}\,\text{J})}{9.11 \times 10^{-31}\,\text{kg}}} = 1.5 \times 10^6\,\text{m/s} \;\text{ and }\; v_e = \sqrt{\dfrac{2(9.6 \times 10^{-19}\,\text{J})}{1.67 \times 10^{-27}\,\text{kg}}} = 3.4 \times 10^4\,\text{m/s}.$$

As expected, the speed of the proton is lower.

3.   Capacitance (Section 16.3)

A **capacitor** consists of two separated conductors and is used to store charge, and electric energy, in the form of an electric field. **Capacitance** (C) is a quantitative measure of how effective a capacitor is in storing charge and is defined as $C = \dfrac{Q}{V}$, where Q is the charge on each conductor (the two conductors have equal but opposite charge), and V is the potential difference between the conductors. The SI unit of capacitance is farad (F).

Note: From now on the symbol V stands for potential *difference* and means the same as ΔV used earlier.

A common capacitor is the parallel-plate capacitor. It consists of two parallel metal plates of area A separated by a distance d. The capacitance of a parallel-plate capacitor is given by $C = \dfrac{\varepsilon_o A}{d}$, where ε_o is the *permittivity of free space* (vacuum) and is equal to $8.85 \times 10^{-12}\,\text{C}^2/(\text{N·m}^2)$. ε_o is a fundamental constant and is related to Coulomb's constant by $k = 1/(4\pi\varepsilon_o) = 9.00 \times 10^9\,\text{N·m}^2/\text{C}^2$.

The energy stored in a capacitor of capacitance C and potential difference V (with a charge $Q = CV$) is given by $U_C = \frac{1}{2}QV = \dfrac{Q^2}{2C} = \frac{1}{2}CV^2$. The form $\frac{1}{2}CV^2$ is usually the most practical, since the capacitance and the applied voltage are often known or can be measured most easily.

Example 16.5   A parallel-plate capacitor consists of plates of area $1.5 \times 10^{-4}\,\text{m}^2$ separated by 2.0 mm. The capacitor is connected to a 12-V battery.

(a) What is the capacitance?

(b) What is the charge on the plates?

(c) How much energy is stored in the capacitor?

(d) What is the electric field between the plates?

Solution: Given: $A = 1.5 \times 10^{-4}$ m^2, $d = 2.0$ mm $= 2.0 \times 10^{-3}$ m, $V = 12$ V.

Find: (a) C (b) Q (c) U_C (d) E.

(a) $C = \dfrac{\varepsilon_0 A}{d} = \dfrac{[8.85 \times 10^{-12} \text{ C}^2/(\text{N·m}^2)](1.5 \times 10^{-4} \text{ m}^2)}{2.0 \times 10^{-3} \text{ m}} = 6.6 \times 10^{-13}$ F $= 0.66$ pF.

(b) $Q = CV = (6.6 \times 10^{-13} \text{ F})(12 \text{ V}) = 7.9 \times 10^{-12}$ C.

(c) $U_C = \frac{1}{2}CV^2 = \frac{1}{2}(6.6 \times 10^{-13} \text{ F})(12 \text{ V})^2 = 4.8 \times 10^{-11}$ J.

(d) $E = \dfrac{\Delta V}{d} = \dfrac{V}{d} = \dfrac{12 \text{ V}}{2.0 \times 10^{-3} \text{ m}} = 6.0 \times 10^3$ V/m.

---

4. Dielectrics (Section 16.4)

A **dielectric** is any nonconducting material (insulator) capable of being partially polarized when placed in an electric field. When inserted between the plates of a capacitor, a dielectric raises the capacitance by a factor κ, the **dielectric constant** (it is always greater than 1), which is defined as $\kappa = \dfrac{C}{C_0}$, where C_0 is the capacitance without the dielectric (or vacuum between the plates). As a result, the energy of the capacitor also increases if the potential difference is maintained constant across the capacitor, $U_C = \frac{1}{2}CV^2 = \frac{1}{2}(\kappa C_0)V^2 = \kappa(\frac{1}{2}CV^2) = \kappa U_0$. The product $\kappa \varepsilon_0$ is called the **dielectric permittivity**, ε (that is, $\varepsilon = \kappa \varepsilon_0$).

For an isolated capacitor (battery disconnected) with charge Q_0 and voltage V_0, the insertion of a dielectric decreases the voltage to V, and the charge remains constant. Therefore, capacitance increases to $C = \dfrac{Q_0}{V}$ $(V < V_0)$. For a capacitor connected to a battery of voltage V_0, the insertion of a dielectric increases the charge Q, and the voltage remains constant. Therefore, capacitance increases to $C = \dfrac{Q}{V_0}$ $(Q > Q_0)$.

Example 16.6 Repeat Example 16.5 if the capacitor is filled with paper. (Assume the capacitor remains connected to the battery.)

Solution: Additional given: $\kappa = 3.7$ (from Table 16.2).

(a) $C = \kappa C_0 = (3.7)(6.6 \times 10^{-13} \text{ F}) = 2.44 \times 10^{-12}$ F $= 2.4$ pF. Note that C increases.

(b) $Q = CV = (2.44 \times 10^{-12} \text{ F})(12 \text{ V}) = 2.9 \times 10^{-11}$ C. Note that Q increases.

(c) $U_C = \frac{1}{2}CV^2 = \frac{1}{2}(2.44 \times 10^{-12} \text{ F})(12 \text{ V})^2 = 1.8 \times 10^{-10}$ J. Note that U_C increases.

(d) $E = \dfrac{\Delta V}{d} = \dfrac{V}{d} = \dfrac{12 \text{ V}}{2.0 \times 10^{-3} \text{ m}} = 6.0 \times 10^3$ V/m. E is the same. Why?

---

5.   Capacitors in Series and Parallel (Section 16.5)

In a series combination, the **equivalent series capacitance** (C_s) is always less than the value of the smallest capacitor in the combination. It is given by $\frac{1}{C_s} = \frac{1}{C_1} + \frac{1}{C_2} + \frac{1}{C_3} + \ldots$. The charge on the equivalent capacitor is the same as that on each individual capacitor ($Q_s = Q_1 = Q_2 = Q_3 = \ldots$), and the voltage across the equivalent capacitor is equal to the sum of the voltages on each capacitor ($V_s = V_1 + V_2 + V_3 + \ldots$).

In a parallel combination, the **equivalent parallel capacitance** (C_p) is always larger than the value of the largest capacitor in the combination. It is given by $C_p = C_1 + C_2 + C_3 + \ldots$. The charge on the equivalent capacitor is equal to the sum of the charges on each capacitor ($Q_p = Q_1 + Q_2 + Q_3 + \ldots$), and the voltage across the equivalent capacitor is the same as that across each individual capacitor ($V_p = V_1 = V_2 = V_3 = \ldots$).

If there are only two capacitors in a series combination, the equivalent series capacitance can be conveniently expressed as $C_s = \frac{1}{1/C_s} = \frac{1}{1/C_1 + 1/C_2} = \frac{C_1 C_2}{C_1 + C_2}$. If the two capacitors have the same capacitance, $C_1 = C_2 = C$, then $C_s = \frac{CC}{C + C} = \frac{C}{2}$.

Note: If you are finding the equivalent capacitance for a series combination, the equivalent capacitance (C_s) should always be smaller than the smallest capacitance in the series combination. If you are finding the equivalent capacitance for a parallel combination, the equivalent capacitance (C_p) should always be greater than the largest capacitance in the parallel combination.

Example 16.7   For the capacitor network shown in the accompanying figure,

$C_1 = 6.00~\mu F$, $C_2 = 8.00~\mu F$, $C_3 = 14.0~\mu F$.

(a) What is the equivalent capacitance?

(b) What is the charge of each capacitor?

(a) What is the voltage across each capacitor?

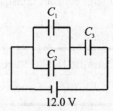

Solution: Given: $C_1 = 6.00~\mu F$, $C_2 = 8.00~\mu F$, $C_3 = 14.0~\mu F$, $V = 12.0$ V.

Find: (a) C_{eq} (b) Q of each capacitor   (c) V across each capacitor.

(a) When working with capacitor combinations, always start with the combination you recognize. In this Example, C_1 and C_3 are *not* in series, and neither are C_2 and C_3; however, C_1 and C_2 are in parallel, so we start from this combination.

$C_p = C_1 + C_2 = 6.00 \ \mu F + 8.00 \ \mu F = 14.0 \ \mu F.$

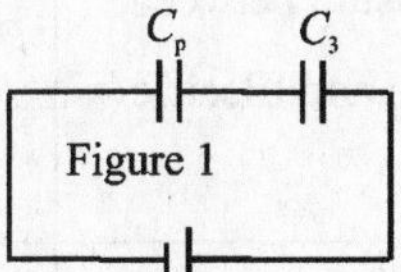

The circuit reduces to (or is equivalent to) Figure 1.

Now we can see that C_p and C_3 are in series:

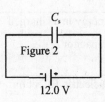

Figure 1

12.0 V

$$\frac{1}{C_s} = \frac{1}{C_p} + \frac{1}{C_3} = \frac{1}{14.0 \ \mu F} + \frac{1}{14.0 \ \mu F} = \frac{2}{14.0 \ \mu F} + \frac{1}{7.00 \ \mu F}.$$

So $C_{eq} = C_s = 7.00 \ \mu F.$

Or we can use $C_s = \dfrac{C_p C_3}{C_p + C_3} = \dfrac{(14.0 \ \mu F)(14.0 \ \mu F)}{14.0 \ \mu F + 14.0 \ \mu F} = 7.00 \ \mu F.$

The circuit becomes Figure 2 and the equivalent capacitance is $7.00 \ \mu F.$

C_s

Figure 2

12.0 V

(b) From Figure 2, the charge on the series capacitor (equivalent capacitor in this example) is $Q_s = C_s V = (7.00 \ \mu F)(12.0 \ V) = 84.0 \ \mu C.$

Because series capacitors have the same charge, C_p and C_3 in Figure 1 will have the same charge as C_s.

So $C_3 = 84.0 \ \mu C$ and $C_p = 84.0 \ \mu C.$ $V_3 = \dfrac{Q_3}{C_3} = \dfrac{84.0 \ \mu C}{14.0 \ \mu F} = 6.00 \ V.$

Because the voltage across the series equivalent capacitor is the sum of the voltages across each individual capacitor, $V_s = V_p + V_3$, so $V_p = V_s - V_3 = 12.0 \ V - 6.00 \ V = 6.00 \ V.$

Parallel capacitors have the same voltage, so $V_1 = V_2 = V_p = 6.00 \ V.$

Therefore, $Q_1 = C_1 V_1 = (6.00 \ \mu F)(6.00 \ V) = 36.0 \ \mu C$ and $Q_2 = C_2 V_2 = (8.00 \ \mu F)(6.00 \ V) = 48.0 \ \mu C.$

Hence, the answers are $Q_1 = 36.0 \ \mu C$, $Q_2 = 48.0 \ \mu C$, and $Q_3 = 84.0 \ \mu C.$

(c) $V_1 = V_2 = V_3 = 6.00 \ V.$

III. Mathematical Summary

Electric Potential Difference (voltage)	$\Delta V = \dfrac{\Delta U_e}{q_+} = \dfrac{W}{q_+}$ (16.1)	Defines the electric potential difference (voltage) in terms of electric potential energy difference.
Electric Potential due to a Point Charge	$V = \dfrac{kq}{r}$ (16.4) $(V = 0 \text{ at } r = \infty)$	Computes the electric potential due to a point charge by assuming zero potential at infinity.
Electrical Potential Energy for a Pair of Point Charges	$U_{12} = \dfrac{kq_1 q_2}{r_{12}}$ (16.5) $(U = 0 \text{ at } r = \infty)$	Calculates the electric potential energy for two point charges.
Electric Potential Energy of Point Charges	$U_{total} = U_{12} + U_{23} + U_{13} + \ldots$ (16.6)	Calculates the electric potential energy of a configuration of point charges.

Relationship between Potential and Electric Field	$E = \left. \dfrac{\Delta V}{\Delta x} \right\|_{max}$ $\quad$ (16.8)	Defines electric field in terms of electric potential gradient (difference).
Capacitance	$Q = CV$ or $C = \dfrac{Q}{V}$ $\quad$ (16.9)	Defines the capacitance of a capacitor.
Capacitance of a Parallel-Plate Capacitor (in air)	$C = \dfrac{\varepsilon_o A}{d}$ $\quad$ (16.12) $\varepsilon_o = 8.85 \times 10^{-12} \ C^2/(N{\cdot}m^2)$	Computes the capacitance of a parallel-plate capacitor.
Energy in a Charged Capacitor	$U_C = \tfrac{1}{2} Q V = \dfrac{Q^2}{2C} = \tfrac{1}{2} C V^2$ $\quad$ (16.13)	Calculates the energy stored in a capacitor.
Dielectric Effect on Capacitance	$C = \kappa C_o$ $\quad$ (16.15)	Calculates the effect of dielectric on capacitance.
Equivalent Series Capacitance	$\dfrac{1}{C_s} = \dfrac{1}{C_1} + \dfrac{1}{C_2} + \dfrac{1}{C_3} + \dots$ $\quad$ (16.17)	Computes the equivalent capacitance for capacitors connected in series.
Equivalent Parallel Capacitance	$C_p = C_1 + C_2 + C_3 + \dots$ $\quad$ (16.18)	Computes the equivalent capacitance for capacitors connected in parallel.

IV. Solutions of Selected Exercises and Paired Exercises

7. It will move to the right or toward the higher potential region because the electron has negative charge. The higher the potential region, for the electron, the lower the potential energy.

14. The electron will accelerate downward because it is negatively charged.

$$a = \frac{F}{m} = \frac{qE}{m}, \quad \text{and} \quad v^2 = v_o^2 + 2ax = 2ax.$$

$$v = \sqrt{2ax} = \sqrt{\frac{2qEx}{m}} = \sqrt{\frac{2(1.6 \times 10^{-19} \ C)(1000 \ V/m)(0.0010 \ m)}{9.11 \times 10^{-31} \ kg}} = \boxed{5.9 \times 10^5 \ m/s \ down}.$$

18. (a) A large orbit will have a $\boxed{(3) \ lower}$ electric potential than a smaller orbit because electric potential is inversely proportional to the distance, $V = \dfrac{kq}{r}$.

(b) $\Delta V = \dfrac{kq}{r_B} - \dfrac{kq}{r_A} = \dfrac{(9.0 \times 10^9 \ N{\cdot}m^2/C^2)(1.6 \times 10^{-19} \ C)}{0.48 \times 10^{-9} \ m} - \dfrac{(9.0 \times 10^9 \ N{\cdot}m^2/C^2)(1.6 \times 10^{-19} \ C)}{0.21 \times 10^{-9} \ m}$

$= -\boxed{3.9 \ V}.$

24.   (a) $W = \Delta U_{A-C} = q\Delta V = qEd = (-1.60 \times 10^{-19}\ C)(15\ V/m)(0.25\ m) = \boxed{-6.0 \times 10^{-19}\ J}$.

(b) $\Delta V_{A-C} = \dfrac{\Delta U_{A-C}}{q} = \dfrac{-6.0 \times 10^{-19}\ J}{-1.60 \times 10^{-19}\ C} = \boxed{3.8\ V}$.

(c) Since ΔV_{A-C} is positive, $\boxed{\text{point C}}$ is at a higher potential.

28.   (a) The distance from the charges to the square center is $r = \sqrt{(0.05\ m)^2 + (0.05\ m)^2} = 0.0707\ m$.

$V = \Sigma\ \dfrac{kq}{r} = 2\ \dfrac{(9.0 \times 10^9\ N{\cdot}m^2/C^2)(-10 \times 10^{-6}\ C)}{0.0707\ m} + 2\ \dfrac{(9.0 \times 10^9\ N{\cdot}m^2/C^2)(5.0 \times 10^{-6}\ C)}{0.0707\ m}$

$= \boxed{-1.3 \times 10^6\ V}$.

(b) The distance from q_2 and q_3 to the point is

$r = \sqrt{(0.10\ m)^2 + (0.05\ m)^2} = 0.112\ m$.

$V = \Sigma\ \dfrac{kq}{r} = \dfrac{(9.0 \times 10^9\ N{\cdot}m^2/C^2)(-10 \times 10^{-6}\ C)}{0.05\ m} + \dfrac{(9.0 \times 10^9\ N{\cdot}m^2/C^2)(-10 \times 10^{-6}\ C)}{0.112\ m}$

$+ \dfrac{(9.0 \times 10^9\ N{\cdot}m^2/C^2)(5.0 \times 10^{-6}\ C)}{0.112\ m} + \dfrac{(9.0 \times 10^9\ N{\cdot}m^2/C^2)(5.0 \times 10^{-6}\ C)}{0.05\ m} = \boxed{-1.3 \times 10^6\ V}$.

29.   (a) The left side of the gun should be at $\boxed{\text{(3) a lower}}$ potential than the right side because electrons have a

negative charge. They move toward higher potential regions where they have lower potential energy.

(b) The change in kinetic energy is $\Delta K = K - K_o = K = \frac{1}{2}mv^2$.

Also from work-energy theorem: $W = \Delta K = -\Delta U_e = -q\Delta V = e\Delta V$.

So $v = \sqrt{\dfrac{2e\Delta V}{m}} = \dfrac{2(1.6 \times 10^{-19}\ C)(5.0 \times 10^3\ V)}{9.11 \times 10^{-31}\ kg} = 4.19 \times 10^7\ m/s = \boxed{4.2 \times 10^7\ m/s}$.

(c) $t = \dfrac{\Delta x}{v} = \dfrac{0.25\ m}{4.19 \times 10^7\ m/s} = \boxed{6.0 \times 10^{-9}\ s}$.

40.   (a) Yes. Electric field is a measure of the change in electric potential over a distance. Inside a conductor in

electrostatic equilibrium, the electric field is zero, yet the electrical potential could be at a *constant* value.

(b) Yes. For example, at the midpoint between a positive and negative charges of equal magnitude, the

potential is zero yet the electric field is not.

42.   Since $E = \dfrac{\Delta V}{\Delta x}$, $\Delta x = \dfrac{\Delta V}{E} = \dfrac{10\ V}{100\ V/m} = 10^{-2}\ m = \boxed{1.0\ cm}$.

48. $W = \Delta K = K - K_o = K = -\Delta U_e = -q\Delta V = e\Delta V = e(100 \times 10^6\ V) = \boxed{1.00 \times 10^8\ eV}$.

$(1.00 \times 10^8\ eV) \times \dfrac{1.60 \times 10^{-19}\ J}{1\ eV} = \boxed{1.60 \times 10^{-11}\ J}$.

52.   (a) $\Delta V = \dfrac{\Delta U_e}{q} = \dfrac{\Delta U_e}{e} = \dfrac{3.5 \text{ eV}}{e} = \boxed{3.5 \text{ V}}$.

From energy conservation, $\frac{1}{2}mv^2 = K = |\Delta U_e|$,

so $v = \sqrt{\dfrac{2|\Delta U_e|}{m}} = \sqrt{\dfrac{2(3.5 \text{ eV})(1.6 \times 10^{-19} \text{ J/eV})}{1.67 \times 10^{-27} \text{ kg}}} = \boxed{2.6 \times 10^4 \text{ m/s}}$.

(b) $\Delta V = \boxed{4.1 \text{ kV}}$. $v = \sqrt{\dfrac{2(4.1 \times 10^3 \text{ eV})(1.6 \times 10^{-19} \text{ J/eV})}{1.67 \times 10^{-27} \text{ kg}}} = \boxed{8.9 \times 10^5 \text{ m/s}}$.

(c) $8.0 \times 10^{-16} \text{ J} = (8.0 \times 10^{-16} \text{ J}) \times \dfrac{1 \text{ eV}}{1.60 \times 10^{-19} \text{ J}} = 5.0 \times 10^3 \text{ eV} = 5.0 \text{ keV}$.

$\Delta V = \boxed{5.0 \text{ kV}}$. $v = \sqrt{\dfrac{2(5.0 \times 10^3 \text{ eV})(1.6 \times 10^{-19} \text{ J/eV})}{1.67 \times 10^{-27} \text{ kg}}} = \boxed{9.8 \times 10^5 \text{ m/s}}$.

63.   (a) Since $Q = CV$, $\boxed{\text{it doubles}}$.

(b) Since $U_C = \frac{1}{2}CV^2$, $\boxed{\text{it quadruples}}$.

66. $C = \dfrac{\varepsilon_o A}{d} = \dfrac{(8.85 \times 10^{-12} \text{ F/m})(0.50 \text{ m}^2)}{2.0 \times 10^{-3} \text{ m}} = \boxed{2.2 \times 10^{-9} \text{ F}}$.

68.   (a) A large plate area results in $\boxed{\text{(1) a larger}}$ capacitance because capacitance is directly proportional to the

plate area, $C = \dfrac{\varepsilon_o A}{d}$.

(b) From $C = \dfrac{\varepsilon_o A}{d}$, we have $\dfrac{A_2}{A_1} = \dfrac{C_2}{C_1}$.

So $A_2 = \dfrac{C_2}{C_1} A_1 = 2(0.425 \text{ m}^2) = \boxed{0.850 \text{ m}^2}$.

82.   (a) From $C = \dfrac{\kappa \varepsilon_o A}{d}$, we have $d = \dfrac{\kappa \varepsilon_o A}{C} = \dfrac{4.6(8.85 \times 10^{-12} \text{ F/m})(0.50 \text{ m}^2)}{0.10 \times 10^{-6} \text{ F}} = \boxed{0.20 \text{ mm}}$.

(b) $Q = CV = (0.10 \times 10^{-6} \text{ F})(12 \text{ V}) = \boxed{1.2 \ \mu\text{C}}$.

90.   (a) Connect them in parallel to get maximum equivalent capacitance.

(b) Connect them in series to get minimum equivalent capacitance.

93. (a) The parallel combination will draw $\boxed{\text{(1) more}}$ energy from the battery because the equivalent capacitance is higher and the energy stored (drawn) is proportional to capacitance.

(b) From $U_{\text{total}} = \frac{1}{2}C_s V^2$, we have $C_s = \frac{2U_{\text{total}}}{V^2} = \frac{2(173\ \mu\text{J})}{(12\ \text{V})^2} = 2.40\ \mu\text{F}$.

Also $\frac{1}{C_s} = \frac{1}{C_1} + \frac{1}{C_2}$, so $C_2 = \frac{C_1 C_s}{C_1 - C_s} = \frac{(4.0\ \mu\text{C})(2.40\ \mu\text{F})}{4.0\ \mu\text{F} - 2.40\ \mu\text{F}} = \boxed{6.0\ \mu\text{F}}$.

96. (a) You can obtain $\boxed{\text{(3) seven}}$ different capacitance values.

(b) Three in series: $\frac{1}{C_s} = \frac{1}{C_1} + \frac{1}{C_2} + \frac{1}{C_2} = \frac{3}{1.0\ \mu\text{F}}$, so $C_s = \boxed{0.33\ \mu\text{F}}$.

Two in series: $\frac{1}{C_s} = \frac{1}{C_1} + \frac{1}{C_2} = \frac{2}{1.0\ \mu\text{F}}$, so $C_s = \boxed{0.50\ \mu\text{F}}$.

Two in parallel, then series: $C_p = C_1 + C_2 = 2.0\ \mu\text{F}$.

So $C_s = \frac{C_p C_3}{C_p + C_3} = \frac{(2.0\ \mu\text{F})(1.0\ \mu\text{F})}{2.0\ \mu\text{F} + 1.0\ \mu\text{F}} = \boxed{0.67\ \mu\text{F}}$.

Just one: $C = \boxed{1.0\ \mu\text{F}}$.

Two in series, then parallel: $C_p = 0.50\ \mu\text{F} + 1.0\ \mu\text{F} = \boxed{1.5\ \mu\text{F}}$.

Two in parallel: $C_p = 1.0\ \mu\text{F} + 1.0\ \mu\text{F} = \boxed{2.0\ \mu\text{F}}$.

Three in parallel: $C_p = 1.0\ \mu\text{F} + 1.0\ \mu\text{F} + 1.0\ \mu\text{F} = \boxed{3.0\ \mu\text{F}}$.

98. The voltage is the same for all capacitors in parallel.

$Q_1 = C_1 V = (0.10\ \mu\text{F})(6.0\ \text{V}) = \boxed{0.60\ \mu\text{C}}$,

$Q_2 = (0.20\ \mu\text{F})(6.0\ \text{V}) = \boxed{1.2\ \mu\text{C}}$, $Q_3 = (0.30\ \mu\text{F})(6.0\ \text{V}) = \boxed{1.8\ \mu\text{C}}$.

106. (a) $W = \Delta U_e = q\Delta V = (1.60 \times 10^{-19}\ \text{C})(30 \times 10^{-3}\ \text{V}) = \boxed{4.8 \times 10^{-21}\ \text{J}} = 0.030\ \text{eV}$.

(b) $E = \frac{V}{d} = \frac{30 \times 10^{-3}\ \text{V}}{10 \times 10^{-9}\ \text{m}} = 3.0 \times 10^6\ \text{V/m} = \boxed{3.0\ \text{MV/m outward}}$, because the exterior is at a lower potential.

(c) $E = \frac{70 \times 10^{-3}\ \text{V}}{10 \times 10^{-9}\ \text{m}} = 7.0 \times 10^6\ \text{V/m} = \boxed{7.0\ \text{MV/m inward}}$, because the exterior is at a higher potential.

V.   Practice Quiz

1.   For an electron moving in the direction of the electric field,

   (a) its potential energy increases, and its potential increases.

   (b) its potential energy decreases, and its potential increases.

   (c) its potential energy increases, and its potential decreases.

   (d) its potential energy decreases, and its potential decreases.

   (e) both the potential energy and potential remain constant.

2.   The electric field between two parallel plates separated by 0.50 cm is 6.0×10^3 V/m. What voltage across the plates is required?

   (a) 3.0 V   (b) 30 V   (c) 300 V   (d) 1.2×10^4 V   (e) 1.2×10^6 V

3.   A proton moves 0.10 m along the direction of an electric field line of magnitude 3.0 V/m. What is the magnitude of the change in potential energy of the proton?

   (a) 4.8×10^{-20} J   (b) 3.2×10^{-20} J   (c) 1.6×10^{-20} J   (d) 8.0×10^{-21} J   (e) zero

4.   If three capacitors, of 1.0, 1.5, and 2.0 μF, are connected in parallel, what is the equivalent capacitance?

   (a) 4.5 μF   (b) 4.0 μF   (c) 2.17 μF   (d) 0.46 μF   (e) 0.67 μF

5.   If a capacitor is charged so that it stores 0.020 J of electric potential energy when connected to a 120-V power source, what is its capacitance?

   (a) 1.4 pF   (b) 3.8 pF   (c) 1.4 μF   (d) 2.8 μF   (e) 0.33 mF

6.   A parallel plate capacitor has a capacitance C. The area of the plates is doubled, and the distance between the plates is halved. What is the new capacitance?

   (a) $C/4$ (b) $C/2$ (c) C (d) $2C$ (e) $4C$

7.   Two charges of 3.00 μC each are located on the ends of a meterstick. Find the electric potential at the center of the meterstick.

   (a) zero   (b) 2.70×10^4 V   (c) 5.40×10^4 V   (d) 1.08×10^5 V   (e) 2.16×10^5 V

8.   A uniform electric field with a magnitude of 500 V/m is directed parallel to the negative x-axis. If the potential at $x = 5.0$ m is 2500 V, what is the potential at $x = 2.0$ m?

   (a) 500 V   (b) 1000 V   (c) 2000 V   (d) 4000 V   (e) 8000 V

9. What is the equivalent capacitance of the combination shown?

(a) 10 μF

(b) 25 μF

(c) 29 μF

(d) 40 μF

(e) 68 μF

10. If the combination in Exercise 9 is connected to a 12-V battery, what is the charge on the 20-μF capacitor?

(a) 60 μC (b) 120 μC (c) 180 μC (d) 240 μC (e) 360 μC

11. A 12-V battery is connected to a parallel-plate capacitor with a plate area of 0.40 m^2 and a plate separation of 2.0 mm. How much energy is stored in the capacitor?

(a) 1.8×10^{-9} J (b) 2.1×10^{-8} J (c) 4.2×10^{-8} J (d) 1.3×10^{-7} J (e) 5.0×10^{-7} J

12. A parallel-plate capacitor is connected to a battery. If a dielectric is inserted between the plates,

(a) the capacitance decreases. (b) the voltage increases.

(c) the voltage decreases. (d) the charge increases.

(e) the charge decreases.

Answers to Practice Quiz:

1.c 2.b 3.a 4.a 5.d 6.e 7.d 8.b 9.a 10.b 11.d 12.d

CHAPTER 17

Electric Current and Resistance

I.   Chapter Objectives

Upon completion of this chapter, you should be able to:

1. introduce the properties of a battery, explain how a battery produces a direct current in a circuit, and learn various circuit symbols for sketching schematic circuit diagrams.

2. define electric current, distinguish between electron flow and conventional current, and explain the concept of drift velocity and electric energy transmission.

3. define electrical resistance and explain what is meant by an ohmic resistor, summarize the factors that determine resistance, and calculate the effect of these factors in simple situations.

4. define electric power, calculate the power delivery of simple electric circuits, and explain joule heating and its significance.

II.   Chapter Summary and Discussion

1.   Batteries and Direct Current (Section 17.1)

A **battery** is a device that converts chemical potential energy into electrical energy. The potential difference across the two terminals, **anode** (+) and **cathode** (−), of a battery (or any dc power supply) when it is not connected to an external circuit is called the **electromotive force** (EMF). The **terminal voltage** (operating voltage) across a battery or power supply is the voltage when it is connected to an external circuit. Terminal voltage is always less than the EMF because of internal resistance of the battery.

Batteries can be connected in series or parallel. When batteries are connected in series, their voltages add ($V = V_1 + V_2 + V_3 + \ldots$). When batteries of the same voltage are connected in parallel, the total voltage is the same as the voltage of each individual battery ($V = V_1 = V_2 = V_3 \ldots$).

2.   Current and Drift Velocity (Section 17.2)

Electric current is the net rate at which charge flows past a given point. The direction of **conventional current** is that in which positive charge would move. In most materials (e.g., metals), the actual current is carried by electrons moving in the direction opposite the conventional current due to the fact that electrons have negative charge. By convention, a battery supplies a conventional current in a circuit, originating from the positive terminal of the battery and terminating at the negative terminal. This current is called **direct current** (dc). (It is actually the electrons flowing from negative terminal to positive terminal.)

A battery or some other voltage source connected to a continuous conducting path forms a **complete circuit**. Quantitatively, the electric current (I) in a wire is defined as the time rate of flow of net charge. If a net charge q passes a given point in a wire in time t at a constant rate, the current is given by $I = \frac{q}{t}$. The SI units of current are coulomb/s (C/s) or ampere (A).

Example 17.1   If 3.0×10^{15} electrons flow through a section of a wire of diameter 2.0 mm in 4.0 s, what is the electric current in the wire?

Solution: Given: $n = 3.0 \times 10^{15}$ electrons, $t = 4.0$ s, $d = 2.0$ mm.
                    Find: I.

The charge in 3.0×10^{15} electrons is $q = ne = (3.0 \times 10^{15})(1.60 \times 10^{-19} \text{ C}) = 4.8 \times 10^{-4}$ C.

Therefore, $I = \frac{q}{t} = \frac{4.8 \times 10^{-4} \text{ C}}{4.0 \text{ s}} = 1.2 \times 10^{-4}$ A $= 0.12$ mA.

The diameter of the wire is not needed in this example, since the current is simply the rate of charge flow.

The electron flow in a metal wire is characterized by an average velocity called the **drift velocity**, which is much smaller than the random velocities of the electrons themselves. The drift velocity is usually very slow (approximately 0.10 cm/s), but the electric field (which is what pushes the charges in the wire) travels down the wire at a speed close to the speed of light (on the order of 10^8 m/s). Hence, the current starts "instantly" in all parts of the circuits.

3.   Resistance and Ohm's Law (Section 17.3)

The electrical **resistance** of a circuit element is defined as the potential difference (voltage) across the element divided by the resulting current, $R = V/I$. The SI unit of resistance is volt/ampere (V/A), or ohm (Ω). A resistor that has constant resistance is (at a given temperature) said to obey **Ohm's law**, or to be ohmic. Ohm's law is usually written as $V = IR$.

Example 17.2   A 20-Ω resistor is connected to a 12-V battery. What is the current through the resistor?

Solution: Given: $R = 20\ \Omega,\quad V = 12\ \mathrm{V}$.

  Find: I.

From Ohm's law, $I = \dfrac{V}{R} = \dfrac{12\ \mathrm{V}}{20\ \Omega} = 0.60\ \mathrm{A}$.

The major factors affecting the resistance of a conductor of uniform cross-section are

  (1) the type of material or the intrinsic resistive properties,

  (2) its length (L),

  (3) its cross-sectional area (A),

  (4) its temperature (T).

The resistive properties of a particular material are characterized by its **resistivity** (ρ), where for a conductor of uniform cross-section, $\rho = \dfrac{RA}{L}$. The SI units of resistivity are $\Omega \cdot \mathrm{m}$. This resistivity is somewhat temperature dependent, and over a small range of temperature change (ΔT), the resistivity changes according to $\rho = \rho_{\mathrm{o}}(1 + \alpha \Delta T)$, where ρ_{o} is a reference resistivity at T_{o} (usually $0°\mathrm{C}$) and α is the **temperature coefficient of resistivity**.

For a conductor of uniform cross-section, we may write the resistance as $R = \dfrac{\rho L}{A}$ and the variation of resistance with temperature as $R = R_{\mathrm{o}}(1 + \alpha \Delta T)$. Most metallic materials have positive α's, and their resistance increases with temperature.

Example 17.3 Calculate the current in a piece of 10.0-m-long 22-gauge (the radius is 0.321 mm) nichrome wire if it is connected to a 12.0-V source. Assume the temperature is 20°C.

Solution: Given: $L = 10.0\ \mathrm{m},\quad r = 0.321\ \mathrm{mm} = 0.321 \times 10^{-3}\ \mathrm{m},\quad \rho = 100 \times 10^{-8}\ \Omega \cdot \mathrm{m}$ (Table 17.1),

  $V = 12.0\ \mathrm{V}$.

  Find: I.

To find the current I, we first need to find the resistance R of the wire:

$$R = \frac{\rho L}{A} = \frac{\rho L}{\pi r^2} = \frac{(100 \times 10^{-8}\ \Omega \cdot \mathrm{m})(10.0\ \mathrm{m})}{\pi (0.321 \times 10^{-3}\ \mathrm{m})^2} = 30.9\ \Omega.$$

From Ohm's law, $I = \dfrac{V}{R} = \dfrac{12.0\ \mathrm{V}}{30.9\ \Omega} = 0.388\ \mathrm{A}$.

Example 17.4   A carbon resistor has a resistance of 16 Ω at a temperature 20°C. What is the resistance if the resistor is heated to a temperature of 100°C?

Solution: Given: $R_o = 16\ \Omega$, $T_o = 20°C$, $T = 100°C$, $\alpha = -5.0 \times 10^{-4}\ C°^{-1}$ (Table 17.1).

Find: R.

$$R = R_o(1 + \alpha\Delta T) = (16\ \Omega)[1 + (-5.0 \times 10^{-4}\ C°^{-1})(100°C - 20°C)] = 15\ \Omega.$$

The resistance of carbon decreases with increasing temperature. When carbon is used with materials with a positive temperature coefficient, resistors can be made that do not vary with temperature. (Why?)

4.   Electric Power (Section 17.4)

Electric power is the time rate at which work is done or electric energy is transferred. The power delivered to a circuit element depends on its resistance and the voltage across it. The SI units of power are J/s or watt (W). Power can be expressed as $P = IV = \dfrac{V^2}{R} = I^2 R$.

The thermal energy expended in a current-carrying resistor is sometimes referred to as **joule heat**, or I^2R ("*I* squared *R*") **losses**.

Integrated Example 17.5

A normal household lightbulb is rated at 100-W for the nominal voltage of 120 V. (a) If the lightbulb is accidentally connected to a 240-V source, the power of the lightbulb is (1) 100 W, (2) 200 W, or (3) 400 W. Explain. (b) Calculate the resistance, current, and power of the lightbulb when it is connected to 240 V. Assume the bulb is ohmic.

(a) Conceptual Reasoning:

Since the bulb is ohmic, its resistance is a constant, that is, its resistance at 120 V is the same as its resistance at 240 V. According to Ohm's law, the current at 240 V is twice as much as at 120 V. Since the voltage is also doubled, the power is going to be 4 times or 400 W at 240 V because $P = IV$.

(b) Quantitative Reasoning and Solution:

Given: $P_1 = 100\ W$, $V_1 = 120\ V$, and $V_2 = 240\ V$.

Find: R_2, I_2, and P_2.

Since the bulb is ohmic, its resistance at 240 V is the same as its resistance at 120 V.

From $P = \dfrac{V^2}{R}$, we have $R_1 = \dfrac{V_1{}^2}{P_1} = \dfrac{(120\ \text{V})^2}{100\ \text{W}} = 144\ \Omega = R_2$.

From Ohm's law, the current at 240 V is $I_2 = \dfrac{V_2}{R_2} = \dfrac{240\ \text{V}}{144\ \Omega} = 15/9\ \text{A} = 1.7\ \text{A}$.

$P_2 = \dfrac{V_2{}^2}{R_2} = \dfrac{(240\ \text{V})^2}{144\ \Omega} = 400\ \text{W}$.

Or we can use $P_2 = I_2 V_2 = (15/9\ \text{A})(240\ \text{V}) = 400\ \text{W}$.

As expected, the power is 4 time or 400 W.

Example 17.6   A computer, including its monitor, is rated at 300 W. Assuming the power company charges $0.10 for each kilowatt-hour (kWh) of electricity used, and the computer is on 8.0 hours per day, estimate the annual cost to operate the computer.

Solution: Given: $P = 300\ \text{W}$, $t = 8.0\ \text{h}$ every day, cost = $0.10 per kWh.

Find: total annual cost.

In one year, the computer is on for $t = (8.0\ \text{h/day})(365\ \text{day}) = 2.93 \times 10^3\ \text{h}$.

The total electrical energy expended is

$E = Pt = (300\ \text{W})(2.93 \times 10^3\ \text{h}) = (0.30\ \text{kW})(2.93 \times 10^3\ \text{h}) = 8.80 \times 10^2\ \text{kWh}$.

So the total cost is ($0.10/kWh)($8.80 \times 10^2\ \text{kWh}) = $88.

III.   Mathematical Summary

Electric Current	$I = \dfrac{q}{t}$ (17.1)	Defines electric current in terms of charge flow.
Electrical Resistance (definition)	$R = \dfrac{V}{I}$ or $V = IR$ (17.2)	Defines electrical resistance.
Ohm's Law	$V = IR$ (R = constant) (17.2)	Relates voltage, current, and resistance.
Resistivity	$R = \rho \dfrac{L}{A}$ (17.3)	Relates resistance to resistivity.
Electric Power	$P = IV = \dfrac{V^2}{R} = I^2 R$ (17.7b)	Computes the electric power delivery to a resistor.

IV. Solutions of Selected Exercises and Paired Exercises

8.   (a)

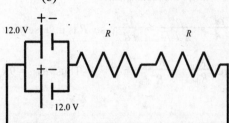

(c)

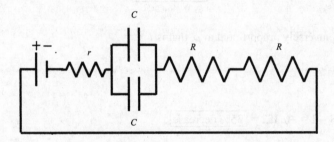

10.   (a) $V = V_1 + V_2 + \ldots V_6 = 6(1.5 \text{ V}) = \boxed{9.0 \text{ V}}$.

(b) $V = V_1 = V_2 = \ldots V_6 = \boxed{1.5 \text{ V}}$.

20.   From $I = \dfrac{q}{t}$, we have $t = \dfrac{q}{I} = \dfrac{2.5 \text{ C}}{5.0 \times 10^{-3} \text{ A}} = 500 \text{ s} = \boxed{8.3 \text{ min}}$.

25.   (a) The net current will be $\boxed{(2) \text{ to the left}}$. The current due to the protons will be to the left, and the

current due to the electrons will also be to the left because electrons have negative charge.

(b) $I = \dfrac{q_{\text{net}}}{t} = \dfrac{-6.7 \text{ C} - (+8.3 \text{ C})}{4.5 \text{ s}} = -\boxed{3.3 \text{ A}}$.

32.   This is because the resistance is low and the current is high at turn on. Once the lamp is hot, its resistance

increases and current decreases, so there is less chance of burning out.

38. $I = \dfrac{V}{R} = \dfrac{12 \text{ V}}{15 \ \Omega} = \boxed{0.80 \text{ A}}$.

44. (a) The resistance of the thinner wire is $\boxed{(3)\ 9\ \text{times}}$ that of the resistance of the thicker wire. Since $R = \dfrac{\rho L}{A}$ and A is proportional to the diameter squared, having 1/3 of the diameter means 1/9 of the area and therefore 9 times the resistance.

(b) $R = \dfrac{\rho L}{A} = \dfrac{\rho L}{\pi d^2/4} = \dfrac{4\rho}{\pi d^2}$, so R is inversely proportional to the square of d, that is, $R \propto \dfrac{1}{d^2}$.

Therefore $\dfrac{R_{\text{thin}}}{R_{\text{thick}}} = \dfrac{R^2_{\text{thick}}}{R^2_{\text{thin}}} = 3^2 = 9$. Thus $R_{\text{thin}} = 9R_{\text{thick}} = 9(1.0\ \Omega) = \boxed{9.0\ \Omega}$.

46. (a) $\rho = \rho_0(1 + \alpha \Delta T) = \rho_0[1 + (6.80 \times 10^{-3}\ \text{C}^{\circ-1})(100°\text{C} - 20°\text{C})] = 1.544\rho_0$.

So the percentage variation is $\dfrac{\rho - \rho_0}{\rho_0} = \dfrac{\rho}{\rho_0} - 1 = 0.544 = \boxed{54\%\ \text{increase}}$.

(b) Since $I = \dfrac{V}{R}$ and $R = \dfrac{\rho L}{A}$, I is inversely proportional to ρ, that is $I \propto \dfrac{1}{\rho}$.

So $\dfrac{I}{I_0} = \dfrac{\rho_0}{\rho} = \dfrac{1}{1.544} = 0.648$.

Therefore $\dfrac{I - I_0}{I_0} = \dfrac{I}{I_0} - 1 = 0.648 - 1 = -0.352 = \boxed{35\%\ \text{decrease}}$.

51. (a) The resistance after the stretch will be $\boxed{(1)\ \text{greater than}}$ that before the stretch. After the stretch, the length L increases and the cross-sectional area A decreases, so R increases according to $R = \dfrac{\rho L}{A}$.

(b) The volume (material) of the wire remains constant and the volume is equal to cross-sectional area times length. $A_1 L_1 = A_2 L_2$, so $\dfrac{A_1}{A_2} = \dfrac{L_2}{L_1}$.

Also $R = \dfrac{\rho L}{A}$, therefore $\dfrac{R_2}{R_1} = \dfrac{L_2}{L_1} \dfrac{A_1}{A_2} = \left(\dfrac{L_2}{L_1}\right)^2 = \left(\dfrac{1.25}{1}\right)^2 = \boxed{1.6}$.

54. (a) The current will $\boxed{(3)\ \text{decrease}}$ if the temperature increases. Platinum has a positive temperature coefficient of resistivity, so its resistivity, and therefore resistance, increases with temperature. The increased resistance causes a decrease in current according to Ohm's law.

(b) $I_0 = \dfrac{V}{R_0} = \dfrac{1.5\ \text{V}}{5.0\ \Omega} = 0.30\ \text{A}$.

$I = \dfrac{V}{R} = \dfrac{V}{R_0(1 + \alpha \Delta T)} = \dfrac{1.5\ \text{V}}{(5.0\ \Omega)[1 + (3.93 \times 10^{-3}\ \text{C}^{\circ-1})(2000\ \text{C}°)]} = 0.0334\ \text{A}$.

So $\Delta I = 0.0334\ \text{A} - 0.30\ \text{A} = -0.27\ \text{A}$, i.e., $\boxed{\text{decrease by } 0.27\ \text{A}}$.

62. $P = \dfrac{V^2}{R} = \dfrac{(110\ \text{V})^2}{10\ \Omega} = \boxed{1.2 \times 10^3\ \text{W}}$.

67. (a) The resistor will dissipate $\boxed{(4)\ 1/4}$ times the designed power if the voltage is halved. If the voltage is halved, the current is also halved. Power is equal to voltage times current, so power becomes 1/4 of its original value.

(b) From $P = IV = \dfrac{V^2}{R}$, we have $\dfrac{P}{P_o} = \dfrac{V^2}{V_o^2} = \dfrac{(40\ \text{V})^2}{(120\ \text{V})^2} = 1/9$. So $P = \dfrac{P_o}{9} = \dfrac{90\ \text{W}}{9} = \boxed{10\ \text{W}}$.

72. $E = Pt = (0.200\ \text{kW})(10\ \text{h/day})(365\ \text{day/year}) = 730\ \text{kWh/year}$.

So the annual cost is $(730\ \text{kWh})(\$0.15/\text{kWh}) = \boxed{\$110}$.

76. (a) The dissipated power will $\boxed{(3)\ \text{decrease}}$ because the current in the wire decreases. As temperature increases, the resistance of the wire increases. Since power is voltage times current, the power decreases.

(b) $R = R_o(1 + \alpha\,\Delta T) = R_o[1 + (4.5 \times 10^{-3}\ \text{C}^{\circ-1})(150\ \text{C}^{\circ})] = 1.68 R_o$.

From $P = \dfrac{V^2}{R}$, we have $\dfrac{P}{P_o} = \dfrac{R_o}{R} = \dfrac{1}{1.68} = 0.595$. So $P = 0.595(500\ \text{W}) = 298\ \text{W}$.

Therefore $\Delta P = 298\ \text{W} - 500\ \text{W} \approx -202\ \text{W}$.

Thus it is a $\boxed{\text{decrease by 202 W}}$.

78. (a) $I = \dfrac{P}{V} = \dfrac{5.5 \times 10^3\ \text{W}}{240\ \text{V}} = 23\ \text{A} > 20\ \text{A}$. So it should have a $\boxed{\text{30-A breaker}}$.

(b) The heat (energy) is $Q = cm\Delta T = [4190\ \text{J/(kg}\cdot\text{C}^{\circ})](55\ \text{gal})(3.785\ \text{kg/gal})(60\ \text{C}^{\circ}) = 5.23 \times 10^7\ \text{J}$.

The energy input is $E = \dfrac{5.23 \times 10^7\ \text{J}}{0.85} = 6.16 \times 10^7\ \text{J}$. $t = \dfrac{E}{P} = \dfrac{6.16 \times 10^7\ \text{J}}{5.5 \times 10^3\ \text{W}} = 1.12 \times 10^4\ \text{s} = \boxed{3.1\ \text{h}}$.

83. $E = \Sigma(Pt) = (5.0\ \text{kW})(0.30)(24\ \text{h/d})(30\ \text{d}) + (0.8\ \text{kW})(0.50\ \text{h}) + (1.2\ \text{kW})(8.0\ \text{h})$

$\qquad + (0.900\ \text{kW})(1/4\ \text{h/d})(30\ \text{d}) + (0.5\ \text{kW})(0.15)(24\ \text{h/d})(30\ \text{d}) + (10.5\ \text{kW})(10\ \text{h})$

$\qquad + (0.1\ \text{kW})(120\ \text{h}) = 1268\ \text{kWh}$.

So it costs $(1268\ \text{kWh})(\$0.12\ /\text{kWh}) = \boxed{\$152}$.

84. (a) Since copper has a positive temperature coefficient of resistance and carbon has a negative one, copper will have $\boxed{(1)\ \text{a higher resistance than}}$ the carbon piece.

(b) From $R = R_o(1 + \alpha\Delta T)$, we have $\dfrac{R_{\text{Cu}}}{R_{\text{C}}} = \dfrac{1 + \alpha_{\text{Cu}}\Delta T}{1 + \alpha_{\text{C}}\Delta T} = \dfrac{1 + (6.80 \times 10^{-3}\ \text{C}^{\circ-1})(10.0\ \text{C}^{\circ})}{1 + (-5.0 \times 10^{-4}\ \text{C}^{\circ-1})(10.0\ \text{C}^{\circ})} = \boxed{1.07}$.

86.   (a) $I = \dfrac{\varepsilon}{R + r}$ or $\varepsilon = I(R + r)$.

$\varepsilon = (2.54\text{ A})(4.52\ \Omega + r)$ Eq (1)

$\varepsilon = (4.98\text{ A})(2.21\ \Omega + r)$ Eq (2)

So $(2.54\text{ A})(4.52\ \Omega + r) = (4.98\text{ A})(2.21\ \Omega + r)$.

Solving for $r = \boxed{0.195\ \Omega}$.

(b) $\varepsilon = (2.54\text{ A})(4.52\ \Omega + r) = (2.54\text{ A})(4.52\ \Omega + 0.195\ \Omega) = \boxed{12.0\text{ V}}$.

(c) $V_1 = I_1 R_1 = (2.54\text{ A})(4.52\ \Omega) = \boxed{11.5\text{ V}}$. $V_2 = (4.98\text{ A})(2.21\ \Omega) = \boxed{11.0\text{ V}}$.

90.   (a) $\Delta Q = nVq = nAxq = (nAx)q$.

(b) $I = \dfrac{\Delta Q}{\Delta t} = nq\ \dfrac{x}{\Delta t}\ A = nq v_d A$.

V.   Practice Quiz

1.   What is the current through a 5.0-Ω resistor if the voltage across it is 10 V?

(a) zero   (b) 0.50 A   (c) 2.0 A   (d) 5.0 A   (e) 50 A

2.   A household lightbulb of 100 W is designed to operate at a voltage of 120 V. What is the current through the bulb?

(a) zero (b) 0.83 A   (c) 1.2 A   (d) 12 A   (e) 144 A

3.   A wire carries a steady current of 1.0 A over a period of 20 s. What total charge passes through the wire in this time interval?

(a) 200 C   (b) 20 C   (c) 2.0 C   (d) 0.20 C   (e) 0.05 C

4.   A nichrome wire has a radius of 0.50 mm and a resistivity of $100 \times 10^{-8}\ \Omega\cdot\text{m}$. If the wire carries a current of 0.50 A, what is the voltage across 1.0 m of the wire?

(a) 0.003 V   (b) 0.32 V   (c) 0.64 V   (d) 1.6 V   (e) 1.9 V

5.   The length of a wire is doubled, and the radius is doubled. By what factor does the resistance change?

(a) four times as large   (b) twice as large   (c) unchanged   (d) half as large   (e) quarter as large

6. During a large power demand, the line voltage is reduced by 5.0%. What is the power output of a light bulb rated at 100 W when the voltage is normal?

(a) 2.5 W   (b) 5.0 W   (c) 10 W   (d) 90 W   (e) 95 W

7. A 12-V battery is connected to a 50-Ω automobile dome light. How many electrons flow through the connecting wire in 1 minute?

(a) 0.24   (b) 1.5×10^{18} (c) 3.8×10^{-20} (d) 3.1×10^{18} (e) 9.0×10^{19}

8. A 1500-W heater is connected to a 120-V source for 2.0 h. How much heat energy is produced?

(a) 1.1×10^7 J   (b) 1.8×10^5 J   (c) 9.0×10^4 J   (d) 3.0×10^3 J   (e) 1.5×10^3 J

9. An Internet server rated at 400 W (computer plus monitor) is left on 24 hours per day. If electricity costs $0.10 per kWh, how much does it cost to run the server annually?

(a) $147   (b) $289   (c) $350   (d) $877   (e) $3500

10. A platinum wire is used to determine the melting point of indium. The resistance of the platinum wire is 2.000 Ω at 20°C and increases to 3.072 Ω just as the indium starts to melt. What is the melting temperature of indium?

(a) 117°C   (b) 137°C   (c) 157°C   (d) 351°C   (e) 731°C

11. An application calls for a 20-m-long aluminum wire to have a current of 10 A when a 0.50-V source is used. What is the diameter of the wire?

(a) 3.6×10^{-6} m   (b) 1.1×10^{-5} m   (c) 9.5×10^{-4} m   (d) 1.9×10^{-3} m   (e) 3.8×10^{-3} m

12. You are given three 1.5-V batteries. The first two batteries are connected in parallel, and then this parallel combination is connected in series with the third battery. What is the total voltage of the three-battery combination?

(a) 0 V   (b) 1.5 V   (c) 3.0 V   (d) 4.5 V   (e) none of the preceding

Answers to Practice Quiz:

1. c 2. b 3. b 4. c 5. d 6. d 7. d 8. a 9. c 10. c 11. e 12. c

CHAPTER 18

Basic Electric Circuits

I.   Chapter Objectives

Upon completion of this chapter, you should be able to:

1.   determine the equivalent resistance of resistors in series, parallel, and series-parallel combinations, and use equivalent resistances to analyze simple circuits.

2.   understand the physical principles that underlie Kirchhoff's circuit rules, and apply these rules in the analysis of actual circuits.

3.   understand the charging and discharging of a capacitor through a resistor, and calculate the current and voltage at specific times during these processes.

4.   understand how galvanometers are used as ammeters and voltmeters, how multirange versions of these devices are constructed and how they are connected to measure current and voltage in real circuits.

5.   understand how household circuits are wired, and the underlying principles that govern electric safety devices.

II.   Chapter Summary and Discussion

1.   Resistors in Series, Parallel, and Series-Parallel Combinations (Section 18.1)

When resistors are connected in *series*, the current is the same for all resistors, $I = I_1 = I_2 = I_3 = \ldots$, and the total voltage is the sum of the voltages of the individual resistors, that is, $V = V_1 + V_2 + V_3 + \ldots = \Sigma V_i = \Sigma I R_i$. The **equivalent series resistance** (R_s) is given by $R_s = R_1 + R_2 + R_3 + \ldots = \Sigma R_i$, or the value of equivalent series resistance is the algebraic sum of the individual resistances. The equivalent series resistance is larger than that of the largest resistor in the series combination.

When resistors are connected in *parallel*, the voltage is the same for all resistors, $V = V_1 = V_2 = V_3 = \ldots$, and the total current is the sum of the currents of the individual resistors, that is, $I = I_1 + I_2 + I_3 + \ldots = \Sigma I_i$. The **equivalent parallel resistance** (R_p) is given by $\dfrac{1}{R_p} = \dfrac{1}{R_1} + \dfrac{1}{R_2} + \dfrac{1}{R_3} + \ldots = \Sigma \dfrac{1}{R_i}$, or the reciprocal of the equivalent parallel resistance is equal to the sum of the reciprocals of the individual resistances. The value of equivalent parallel resistance is less than that of the smallest resistor in the parallel combination.

If there are only two resistors in parallel, $\frac{1}{R_p} = \frac{1}{R_1} + \frac{1}{R_2} = \frac{R_1 + R_2}{R_1 R_2}$, so $R_p = \frac{R_1 R_2}{R_1 + R_2}$. If $R_1 = R_2 = R$,

then $R_p = \frac{RR}{R + R} = \frac{R}{2}$.

Note: In Chapter 16, the equivalent series capacitance was given by $\frac{1}{C_s} = \frac{1}{C_1} + \frac{1}{C_2} + \frac{1}{C_3} + \ldots = \Sigma \frac{1}{C_i}$;

the equivalent parallel capacitance is equal to $C_p = C_1 + C_2 + C_3 + \ldots = \Sigma C_i$. Do not confuse resistor combination with capacitor combination.

Example 18.1 In the circuit shown, find:

(a) the equivalent resistance between points A and B.

(b) the current through each resistor.

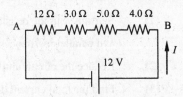

Solution: Given: $R_1 = 12 \, \Omega$, $R_2 = 3.0 \, \Omega$, $R_3 = 5.0 \, \Omega$, $R_4 = 4.0 \, \Omega$, $V = 12$ V.

Find: (a) R_s (b) I.

(a) All four resistors are in series combination, so

$R_s = R_1 + R_2 + R_3 + R_4 = 12 \, \Omega + 3.0 \, \Omega + 5.0 \, \Omega + 4.0 \, \Omega = 24 \, \Omega$.

(b) The current through all resistors in series is the same: $I = \frac{V}{R} = \frac{V}{R_4} = \frac{12 \, \text{V}}{24 \, \Omega} = 0.50$ A.

Example 18.2 In the circuit shown, find:

(a) the equivalent resistance between points A and B.

(b) the current through the battery.

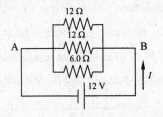

Solution: Given: $R_1 = 12 \, \Omega$, $R_2 = 12 \, \Omega$, $R_3 = 6.0 \, \Omega$, $V = 12$ V.

Find: (a) R_p (b) I.

(a) The three resistors are in a parallel combination:

$\frac{1}{R_p} = \frac{1}{R_1} + \frac{1}{R_2} + \frac{1}{R_3} = \frac{1}{12 \, \Omega} + \frac{1}{12 \, \Omega} + \frac{1}{6.0 \, \Omega} = \frac{1}{12 \, \Omega} + \frac{1}{12 \, \Omega} + \frac{2}{12 \, \Omega} = \frac{4}{12 \, \Omega} = \frac{1}{3.0 \, \Omega}$,

so $R_p = 3.0 \, \Omega$.

Or using $R_p = \dfrac{RR}{R+R} = \dfrac{R}{2}$ for the two 12-Ω resistors in parallel, we have $R_{p1} = \dfrac{12\,\Omega}{2} = 6.0\,\Omega$.

Now the two 6.0-Ω resistors are in parallel, and the equivalent resistance is $R_p = \dfrac{6.0\,\Omega}{2} = 3.0\,\Omega$.

(b) From Ohm's law, $I = \dfrac{V}{R} = \dfrac{V}{R_p} = \dfrac{12\,V}{3.0\,\Omega} = 4.0$ A.

Resistors in a circuit may be connected in a variety of series-parallel combinations. The general procedure for analyzing circuits with different series-parallel combinations of resistors is to find the voltage across and the current through the various resistors as follows:

(1) Starting with the resistor combination farthest from the voltage source, find the equivalent series and parallel resistances.

(2) Reduce the circuit until there is a single loop with one total equivalent resistance.

(3) Find the total current delivered to the reduced circuit using $I = V/R$, where R is the total equivalent resistance.

(4) Expand the reduced circuit in reverse order to find the currents and voltages for the resistors in each step.

Example 18.3   Find the equivalent resistance between points

(a) A and B.

(b) A and C.

(c) B and C.

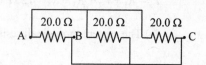

Solution: Given: $R_1 = R_2 = R_3 = R = 20.0\,\Omega$.

Find: (a) R_{AB} (b) R_{AC} (c) R_{BC}.

(a) All three resistors are in parallel between points A and B; that is, the ends of all three resistors are across points A and B. We redraw the circuit.

$$\frac{1}{R_{AB}} = \frac{1}{R_1} + \frac{1}{R_2} + \frac{1}{R_3} = \frac{3}{R},$$

or $R_{AB} = \dfrac{R}{3} = \dfrac{20.0\,\Omega}{3} = 6.67\,\Omega$.

(b) Because points C and B are connected (they are the same point), $R_{AB} = R_{AC} = 6.67\,\Omega$.

(c) Again, since points B and C are connected, there is no resistance between them (short circuit), so $R_{BC} = 0$.

Example 18.4   In the circuit shown, find:

(a) the equivalent resistance between points A and B.

(b) the current through the battery.

(c) the current though the 4.0-Ω resistor.

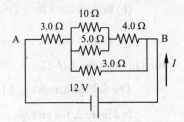

Solution:

Here we have a variety of series-parallel combinations. We follow the four general procedures outlined previously (page 229).

(a) The 10-Ω and the 5.0-Ω resistors are in parallel.

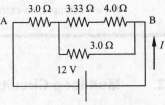

$$\frac{1}{R_p} = \frac{1}{R_1} + \frac{1}{R_2} = \frac{R_1 + R_2}{R_1 R_2}, \text{ so } R_p = \frac{R_1 R_2}{R_1 + R_2}.$$

Therefore, $R_{p1} = \dfrac{(10 \text{ }\Omega)(5.0 \text{ }\Omega)}{10 \text{ }\Omega + 5.0 \text{ }\Omega} = 3.33 \text{ }\Omega.$

The circuit reduces to Figure 1.

Figure 1

Now the 3.33-Ω and the 4.0-Ω resistors are in series:

$R_{s1} = 3.33 \text{ }\Omega + 4.0 \text{ }\Omega = 7.33 \text{ }\Omega.$

The circuit reduces to Figure 2.

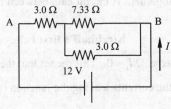

Figure 2

The 7.33-Ω and the 3.0-Ω resistors are in parallel:

$R_{p2} = \dfrac{(7.33 \text{ }\Omega)(3.0 \text{ }\Omega)}{7.33 \text{ }\Omega + 3.0 \text{ }\Omega} = 2.13 \text{ }\Omega.$

The circuit reduces to Figure 3.

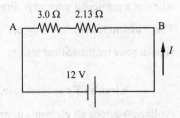

Figure 3

Finally, the 2.13-Ω and the 3.0-Ω resistors are in series.

$R = R_{s2} = 2.13 \text{ }\Omega + 3.0 \text{ }\Omega = 5.13 \text{ }\Omega = 5.1 \text{ }\Omega$ (2 significant figures).

The circuit reduces to Figure 4.

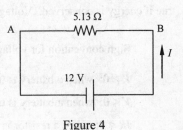

Figure 4

(b) From Ohm's law, $I = \dfrac{V}{R} = \dfrac{12 \text{ V}}{5.13 \text{ }\Omega} = 2.33$ A = 2.3 A (2 significant figures).

(c) To find the current through the 4.0-Ω resistor, we need to expand the combinations. In Figure 3, the current through the 2.13-Ω and 3.0-Ω resistors is the same as the total current, 2.33 A.

The voltage across the 2.13-Ω resistor is then $V_{2.13} = (2.13 \text{ }\Omega)(2.33 \text{ A}) = 4.96$ V.

In Figure 2, the voltage across the 7.33-Ω and 3.0-Ω resistors is the same as that across the 2.13-Ω resistor, 4.96 V. Therefore, the current through the 7.33-Ω resistor is $I_{7.33} = \dfrac{4.96 \text{ V}}{7.33 \text{ }\Omega} = 0.677$ A.

In Figure 1, the current through the 3.33-Ω and the 4.0-Ω resistors is the same as the current through the 7.33-Ω resistor. Therefore, $I_{4.0} = 0.68$ A (2 significant figures).

In this example, an extra digit was carried in intermediate results.

---

2. Multiloop Circuits and Kirchhoff's Rules (Section 18.2)

In a *multiloop circuit*, a **junction** or **node** is a point at which three or more connecting wires are joined together. A circuit path between two junctions is called a **branch** and may contain one or more circuit elements.

Kirchhoff's first rule, or **junction theorem**, states that the algebraic sum of the currents at any junction is zero: $\Sigma I_i = 0$. This means that the sum of the currents going into a junction (taken as positive) is equal to the sum of the currents leaving the junction (taken as negative).

Note: You have to *assume* current directions at a particular junction because you generally cannot tell whether a particular current is directed into or out of a junction simply by looking at a multiloop circuit diagram. These assumptions are totally arbitrary. If these assumptions are wrong, you will find out later from the negative sign in your mathematical results. A negative current will indicate the wrong choice of direction.

Kirchhoff's second rule, or **loop theorem**, states that the algebraic sum of the potential differences (voltages) across all elements of any *closed* loop is zero: $\Sigma V_i = 0$. This means that the sum of the voltage rises is equal to the sum of the voltage drops across the voltage sources and resistors around a closed loop, which must be true if energy is conserved. Voltage drops may be positive or negative, and we will use the following convention.

Sign convention for voltages across circuit elements in traversing a loop:

$V > 0$: when a battery is traversed from its negative terminal to its positive terminal (voltage rises)

$V < 0$: when a battery is traversed from its positive terminal to negative terminal (voltage drops)

$IR < 0$: when a resistor is traversed in the direction of the assigned branch current (voltage drop)

$IR > 0$: when a resistor is traversed in the direction opposite the assigned branch current (voltage rise)

The general steps in applying Kirchhoff's rules are as follows:

(1)   Assign a current and current direction for each branch in the circuit. This is done most conveniently at junctions.

(2)   Indicate the loops and the arbitrarily chosen directions in which they are to be traversed. Every branch *must* be in at least one loop.

(3)   Apply Kirchhoff's first rule and write equations for the currents, one for each junction that gives a different equation. (In general, this gives a set of equations that includes all branch currents.)

(4)   Traverse the number of loops necessary to include all branches. In traversing a loop, apply Kirchhoff's second rule and write equations using the adopted sign convention.

(5)   Solve the simultaneous equations for the currents.

These steps may seem complicated but really are straightforward, as the following examples show; however, if there are more than two simultaneous equations involved, the mathematical solution can be lengthy. With good calculators (for example, Texas Instruments series TI-85), you can solve the simultaneous equations by simply entering the numbers. Ask your instructor if you are not sure how to use this feature of your calculator.

Integrated Example 18.5

Refer to the accompanying circuit. (a) The current in the 4.0-Ω resistor is (1) greater then, (2) the same as, or (3) less than the current in the 5.0-Ω resistor. Explain. (b) Calculate the currents in all three resistors.

(a) Conceptual Reasoning:

First, we assign junctions and currents as shown in the circuit. This is a simple circuit that contains only one battery. We can conclude that the current must be going from b, through the three resistors, and then to a. Therefore for junction c, the current in the 4.0-Ω resistor is to the right, and the currents in the 5.0-Ω and 9.0-Ω resistors are also to the right.

From Kirchhoff's junction rule, we know that the current in the 4.0-Ω resistor is equal to the sum of the current in the 5.0-Ω and 9.0-Ω resistors. Thus, the current in the 4.0-Ω resistor is (1) greater then the current in the 5.0-Ω resistor.

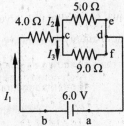

(b) Quantitative Reasoning and Solution:

From the junction rule:

for junction c, $I_1 = I_2 + I_3$.   (1)

From the loop rule:

for loop abceda, $+6.0 \text{ V} - I_1(4.0 \text{ }\Omega) - I_2(5.0 \text{ }\Omega) = 0$,

or $4.0I_1 + 5.0I_2 = 6.0$;   (2)

for loop abcfda, $+6.0 \text{ V} - I_1(4.0 \text{ }\Omega) - I_3(9.0 \text{ }\Omega) = 0$,

or $4.0I_1 + 9.0I_3 = 6.0$.   (3)

From Eq. (1), $I_3 = I_1 - I_2$. Substituting this result into Eq. (3) to eliminate I_3,

$13I_1 - 9.0I_2 = 6.0$.   (4)

From Eq. (2), $I_1 = 1.5 - 1.25I_2$. Substituting this result into Eq. (4) to eliminate I_1, $25.25I_2 = 13.5$.

Therefore, $I_2 = 0.535 \text{ A}$, $I_1 = 1.5 - 1.25I_2 = 1.5 \text{ A} - (1.25)(0.535 \text{ A}) = 0.832 \text{ A}$,

and $I_3 = I_1 - I_2 = 0.832 \text{ A} - 0.535 \text{ A} = 0.297 \text{ A}$.

Hence, $I_1 = 0.83 \text{ A}$ (right), $I_2 = 0.53 \text{ A}$ (right), and $I_3 = 0.30 \text{ A}$ (right), to two significant figures.

BTW, this circuit can also be analyzed with resistor combinations and Ohm's law.

Example 18.6   Find the currents in the branches of the circuit shown.

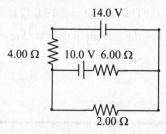

Solution:

You cannot solve this circuit with resistor combinations and Ohm's law. (Why?) You must use Kirchhoff's rules. First, we assign junctions and currents as shown in the circuit.

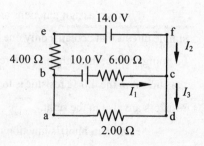

From the junction rule:

for junction c, $I_1 + I_2 = I_3$.   (1)

From the junction rule:

for loop bcfeb, $+10.0 \text{ V} - (6.00 \text{ }\Omega)I_1 + 14.0 \text{ V} + (4.00 \text{ }\Omega)I_2 = 0$,

or $(6.00)I_1 - (4.00)I_2 = 24.0 \text{ V}$;   (2)

for loop bcdab, $+10.0 \text{ V} - (6.00 \ \Omega)I_1 - (2.00 \ \Omega)I_3 = 0,$

or $(6.00)I_1 + (2.00)I_3 = 10.0 \text{ V}.$ (3)

Substituting Eq. (1) into Eq. (3) to eliminate I_3,

$$4I_1 + I_2 = 5.00.$$ (4)

From Eq. (2), $I_2 = 1.50I_1 - 6.00$. Substituting this result into Eq. (4) to eliminate I_2, $11I_1 = 22.0$.

Thus, $I_1 = 2.00 \text{ A}$, $I_2 = 1.50I_1 - 6.00 \text{ A} = -3.00 \text{ A}$, and $I_3 = I_1 + I_2 = 2.00 \text{ A} + (-3.00 \text{ A}) = -1.00 \text{ A}$.

I_2 and I_3 are negative, meaning our assumption (downward) is wrong. These currents should be upward.

Hence the currents are $I_1 = 2.00 \text{ A}$ (right), $I_2 = 3.00 \text{ A}$ (up), and $I_3 = 1.00 \text{ A}$ (up).

---

3.   RC Circuits (Section 18.3)

An **RC circuit** consists of a resistor and a capacitor connected in series. The **time constant** ($\tau = RC$) for an RC circuit is a "characteristic time" by which we measure the capacitor's charge and discharge. In one time constant, $t = \tau$, an uncharged capacitor charges to 63% of its maximum value when charging, and a fully charged capacitor discharges 63% of its charge when discharging (or its charge decreases to 37% of its original value). In charging, both the voltage and the charge on the capacitor increase; in discharging, the voltage and the charge on the capacitor decrease. The current in the circuit always decreases with time regardless of charging or discharging. (Why?) The voltages and currents as functions of time in a capacitor are given by

$V_C = V_0\left(1 - e^{-t/(RC)}\right)$ voltage across capacitor when charging,

$V_C = V_0\,e^{-t/(RC)}$ voltage across capacitor when discharging,

$I = I_0\,e^{-t/(RC)}$ current in the circuit when charging or discharging,

where V_0 and I_0 are the initial voltage across the capacitor and the current in the circuit, respectively, and $\tau = RC$ is the time constant of the circuit.

Example 18.7   In the circuit shown, $C = 5.0 \ \mu\text{F}$, $R = 4.0 \ \text{M}\Omega$, and $V = 12$ V. The switch S is open when the capacitor is initially uncharged. (a) Immediately after the switch is closed, find (1) the voltage across the resistor. (2) the current through the resistor. (b) What is the current in the circuit after 10 s? (c) After the switch has been closed for a long period of time, find (1) the current through the resistor, (2) the charge and energy stored in the capacitor.

Solution: Given: $R = 4.0 \text{ M}\Omega = 4.0 \times 10^6 \ \Omega$, $C = 5.0 \ \mu\text{F} = 5.0 \times 10^{-6} \text{ F}$, $V = 12 \text{ V}$.

(a) (1) Immediately after the switch is closed, the capacitor is not charged, so the voltage across the capacitor is zero ($V_C = 0$). Therefore, the voltage across the resistor is the battery voltage. $V_R = V = 12 \text{ V}$.

(2) $I_o = \dfrac{V}{R} = \dfrac{12 \text{ V}}{4.0 \times 10^6 \ \Omega} = 3.0 \times 10^{-6} \text{ A}$.

(b) The time constant $\tau = RC = (4.0 \times 10^6 \ \Omega)(5.0 \times 10^{-6} \text{ F}) = 20 \text{ s}$, so $\dfrac{t}{RC} = \dfrac{10 \text{ s}}{20 \text{ s}} = 0.50$.

Therefore, $I = I_o e^{-t/(RC)} = (5.0 \times 10^{-6} \text{ A}) \, e^{-0.50} = 3.0 \times 10^{-6} \text{ A}$.   ($e^{-0.50} \approx 0.607$)

(c) (1) After a long period of time, the capacitor is fully charged, and its voltage is $V_C = V = 12 \text{ V}$, so the voltage across the resistor is zero. Therefore, the current through the resistor is zero.

(2) The charge on the capacitor is $Q = CV_C = (5.0 \times 10^{-6} \text{ F})(12 \text{ V}) = 6.0 \times 10^{-5} \text{ C}$.

The energy stored in the capacitor is $U_e = \frac{1}{2}CV^2 = \frac{1}{2}(5.0 \times 10^{-6} \text{ F})(12 \text{ V})^2 = 3.6 \times 10^{-4} \text{ J}$.

---

4.   Ammeters and Voltmeters (Section 18.4)

A **galvanometer** is a sensitive meter with an internal resistance r, whose needle deflection is proportional to the current through it. Only a very small maximum current I_g can exist in a galvanometer without destroying it.

An **ammeter** is a low-resistance device to measure current; it consists of a shunt resistor in parallel with a galvanometer. An ammeter *must* be connected in series with the resistor whose current you want to measure. Because the ammeter has a low resistance, it does not appreciably affect the current measurement. A shunt resistor (R_s) of proper value needs to be selected in constructing an ammeter. The equation that allows us to calculate R_s is

$I_g = \dfrac{IR_s}{r + R_s}$, where I_g is the maximum current through the galvanometer (or the current at full needle deflection), r is the internal resistance of the galvanometer, R_s is the shunt resistance, and I is the current through the ammeter.

A **voltmeter** is a high-resistance device used to measure voltage. It consists of a galvanometer and a multiplier resistor in series with it. A voltmeter *must* be connected in parallel with a circuit element whose voltage you wish to measure. Because the voltmeter has a high resistance, it does not appreciably affect the voltage across an element. A multiplier resistor of proper value must be selected in constructing a voltmeter. The equation that allows us to calculate R_m is $I_g = \dfrac{V}{r + R_m}$, where I_g is the maximum current through the galvanometer, r is the internal resistance of the galvanometer, R_m is the multiplier resistance, and V is the voltage of the voltmeter.

Example 18.8 A galvanometer has an internal resistance 25 Ω and can safely carry 150 μA of current.

(a) What shunt resistor is needed to make an ammeter with a range of 0 to 10 A?

(b) What multiplier resistor is needed to make a voltmeter with a range of 0 to 100 V?

Solution: Given: $r = 25\ \Omega$, $I_g = 150\ \mu A = 150 \times 10^{-6}$ A, $I = 10$ A, $V = 100$ V.

Find: (a) R_s (b) R_m.

(a) From $I_g = \dfrac{IR_s}{r + R_s}$, $R_s = \dfrac{I_g r}{I - I_g} = \dfrac{(150 \times 10^{-6}\ A)(25\ \Omega)}{10\ A - 150 \times 10^{-6}\ A} = 3.8 \times 10^{-4}\ \Omega = 0.38\ m\Omega.$

(b) From $I_g = \dfrac{V}{r + R_m}$, $R_m = \dfrac{V - I_g r}{I_g} = \dfrac{100\ V - (150 \times 10^{-6}\ A)(25\ \Omega)}{150 \times 10^{-6}\ A} = 6.7 \times 10^{5}\ \Omega = 670\ k\Omega.$

5. Household Circuits and Electrical Safety (Section 18.5)

Fuses and **circuit breakers** are safety devices that open (or "break") a circuit when the current exceeds a preset value. If the current in a circuit exceeds a preset (safe) value, this is usually an indication of some problem (short circuit) in the circuit. A **grounded plug** (three-prong plug) uses a dedicated grounding wire to ground objects that may become conductors and thus dangerous. A **polarized plug** identifies the ground side of the line for use as a grounding safety feature. The first precaution to take for personal safety is to avoid coming into contact with an electrical conductor that might cause a voltage across a human body or part of it, thus causing a current through the body that could be dangerous.

III. Mathematical Summary

Equivalent Series Resistance	$R_s = R_1 + R_2 + R_3 + \ldots = \Sigma R_i$ (18.2)	Computes the equivalent series resistance of resistors in series combination.
Equivalent Parallel Resistance	$\dfrac{1}{R_p} = \dfrac{1}{R_1} + \dfrac{1}{R_2} + \dfrac{1}{R_3} + \ldots = \Sigma \dfrac{1}{R_i}$ (18.3)	Computes the equivalent parallel resistance of resistors in parallel combination.
Kirchhoff's Junction Theorem	$\Sigma I_i = 0$ (18.4)	Applies the conservation of charge in electric circuits; sum of currents at junction.
Kirchhoff's Loop Theorem	$\Sigma V_i = 0$ (18.5)	Applies the conservation of energy in electric circuits; sum of voltages around a closed loop.
Time Constant for an RC Circuit	$\tau = RC$ (18.8)	Defines the time constant in a resistor-capacitor series circuit.

IV. Solutions of Selected Exercises and Paired Exercises

10. The $\boxed{40 \text{ W}}$ glows the brightest. The 60 W has a smaller resistance than the 40 W ($P = V^2/R$). When they are connected in series, the larger resistor will consume more power because the current is the same in either bulb ($P = I^2 R$).

    If you switched the order of the bulbs, there is $\boxed{\text{no difference}}$ because order does not matter in series resistors.

    $\boxed{\text{No, neither is at full power rating}}$. At full power rating, the voltage must be 120 V for each bulb. In a series combination, the total voltage is 120 V. That means, either bulb is getting less than 120 V.

14. Series combination: $\quad R_s = R_1 + R_2 = R + R = 2R$.

    Parallel combination: $\quad \dfrac{1}{R_p} = \dfrac{1}{R_s} + \dfrac{1}{R_3}$, so $R_p = \dfrac{R_s R_3}{R_s + R_3} = 10 \, \Omega = \dfrac{(2R)(20 \, \Omega)}{2R + 20 \, \Omega}$.

    Simplifying, $20R + 200 = 40R$. Solving, $R = \boxed{10 \, \Omega}$.

20. (a) The minimum required is $\boxed{(2) \text{ three}}$ resistors.

    (b) $\boxed{\text{Two in parallel, connected in series to the third}}$.

    $R = \dfrac{1.0 \, \Omega}{2} + 1.0 \, \Omega = 0.5 \, \Omega + 1.0 \, \Omega = 1.5 \, \Omega$.

26. R_1 and R_2 are in series: $\quad R_s = 2.0 \, \Omega + 2.0 \, \Omega = 4.0 \, \Omega$.

    R_s, R_3, and R_4 are in parallel: $\quad \dfrac{1}{R_p} = \dfrac{1}{4.0 \, \Omega} + \dfrac{1}{2.0 \, \Omega} + \dfrac{1}{2.0 \, \Omega} = \dfrac{5}{4.0 \, \Omega}$.

    So $R_p = \dfrac{4.0 \, \Omega}{5} = \boxed{0.80 \, \Omega}$.

27. R_2 and R_3 are in series: $\quad R_s = 6.0 \, \Omega + 4.0 \, \Omega = 10 \, \Omega$.

    R_s, R_1, and R_4 are in parallel: $\quad \dfrac{1}{R_p} = \dfrac{1}{10 \, \Omega} + \dfrac{1}{6.0 \, \Omega} + \dfrac{1}{10 \, \Omega} = \dfrac{22}{60 \, \Omega}$.

    So $R_p = \boxed{2.7 \, \Omega}$.

30. The equivalent resistance is

$$R_s = R_p + R_3 = \frac{R_1 R_2}{R_1 + R_2} + R_3 = \frac{(10\ \Omega)(2.0\ \Omega)}{10\ \Omega + 2.0\ \Omega} + 5.0\ \Omega = 6.67\ \Omega.$$

The total current is $I = \dfrac{V}{R_s} = \dfrac{10\ \text{V}}{6.67\ \Omega} = 1.5\ \text{A}.$

The voltage drop across R_3 is $V_3 = (1.5\ \text{A})(5.0\ \Omega) = 7.5\ \text{V}.$

The voltage drop across the $10\ \Omega$ is $10\ \text{V} - 7.5\ \text{V} = \boxed{2.5\ \text{V}}$, and the current is $\dfrac{2.5\ \text{V}}{10\ \Omega} = \boxed{0.25\ \text{A}}.$

36. (a) $I_1 = \dfrac{V}{R_1} = \dfrac{6.0\ \text{V}}{6.0\ \Omega} = \boxed{1.0\ \text{A}}$ R_2 and R_3 are in series. $R_s = 4.0\ \Omega + 6.0\ \Omega = 10\ \Omega.$

So $I_2 = I_3 = \dfrac{6.0\ \text{V}}{10\ \Omega} = \boxed{0.60\ \text{A}}$, $I_4 = \dfrac{6.0\ \text{V}}{10\ \Omega} = \boxed{0.60\ \text{A}}.$

(b) $P_1 = I_1^2 R_1 = (1.0\ \text{A})^2 (6.0\ \Omega) = \boxed{6.0\ \text{W}}$, $P_2 = (0.60\ \text{A})^2 (4.0\ \Omega) = \boxed{1.4\ \text{W}}$,

$P_3 = (0.60\ \text{A})^2 (6.0\ \Omega) = \boxed{2.2\ \text{W}}$, $P_4 = (0.60\ \text{A})^2 (10\ \Omega) = \boxed{3.6\ \text{W}}.$

(c) $P_{\text{sum}} = 6.0\ \text{W} + 1.44\ \text{W} + 2.16\ \text{W} + 3.6\ \text{W} = 13\ \text{W}.$ From Exercise 18.27,

$P_{\text{total}} = \dfrac{(6.0\ \text{V})^2}{2.7\ \Omega} = 13\ \text{W}.$ Therefore $\boxed{P_{\text{sum}} = P_{\text{total}} = 13\ \text{W}}.$

40. $6.0\ \Omega$ and $4.0\ \Omega$ are in parallel. $R_{p1} = \dfrac{(6.0\ \Omega)(4.0\ \Omega)}{6.0\ \Omega + 4.0\ \Omega} = 2.4\ \Omega.$

This R_{p1} and $2.0\ \Omega$ are in series. $R_{s1} = 2.4\ \Omega + 2.0\ \Omega = 4.4\ \Omega.$

This R_{s1} and $12\ \Omega$ are in parallel. $R_{p2} = \dfrac{(4.4\ \Omega)(12\ \Omega)}{4.4\ \Omega + 12\ \Omega} = 3.22\ \Omega.$

$10\ \Omega$ (the one on the bottom) and $5.0\ \Omega$ are in parallel. $R_{p3} = \dfrac{(10\ \Omega)(5.0\ \Omega)}{10\ \Omega + 5.0\ \Omega} = 3.33\ \Omega.$

Finally, R_{p2}, R_{p3}, and $10\ \Omega$ (the one on the top) are in series, $R_{s2} = 3.22\ \Omega + 3.33\ \Omega + 10\ \Omega = 16.6\ \Omega.$

So $P = \dfrac{V^2}{R_{s2}} = \dfrac{(24\ \text{V})^2}{16.6\ \Omega} = \boxed{35\ \text{W}}.$

52. Around the inner loop: $10\ \text{V} - I_2(2.0\ \Omega) - I_3(5.0\ \Omega) = 0.$ Eq. (1)

Around the outer loop: $10\ \text{V} - I_1(10\ \Omega) - I_3(5.0\ \Omega) = 0.$ Eq. (2)

From junction theorem: $I_3 = I_1 + I_2.$ Eq. (3)

Substituting Eq. (3) into Eq. (1) and Eq. (2) gives

$(5.0\ \Omega)I_1 + (7.0\ \Omega)I_2 = 10\ \text{V}$ Eq. (3)

$(7.0\ \Omega)I_1 + (5.0\ \Omega)I_2 = 10\ \text{V}$ Eq. (4)

Solving, $I_1 = \boxed{0.25\ \text{A}}$, $I_2 = \boxed{1.25\ \text{A}}$, and $I_3 = \boxed{1.50\ \text{A}}.$

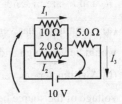

56. (a) For the R_1 and R_2 connecting junction: $I = I_1 + I_2$.   Eq. (1)

Around the loop through R_1 in counterclockwise direction,

$$12 \text{ V} - I(2.0 \ \Omega) - I(8.0 \ \Omega) + 6.0 \text{ V} - I(2.0 \ \Omega) - I_1(4.0 \ \Omega) = 0,$$

$$\text{or} \quad -12I - 4I_1 + 18 = 0. \quad\quad \text{Eq. (2)}$$

Around the loop through R_2 in counterclockwise direction,

$$12 \text{ V} - I(2.0 \ \Omega) - I(8.0 \ \Omega) + 6.0 \text{ V} - I(2.0 \ \Omega) - I_2(6.0 \ \Omega) = 0,$$

$$\text{or} \quad -12I - 6I_1 + 18 = 0. \quad\quad \text{Eq. (3)}$$

Substituting Eq. (1) into Eq. (2) and Eq. (3) gives $-16I_1 - 12I_2 + 18 = 0.$ Eq. (4)

$$-12I_1 - 18I_2 + 18 = 0. \quad \text{Eq. (5)}$$

Solving, $\boxed{I_1 = 0.75 \text{ A left}}$, $\boxed{I_2 = 0.50 \text{ A left}}$, $I = I_3 = I_4 = I_5$,

so $\boxed{I_3 = 1.25 \text{ A up}}$, $\boxed{I_4 = 1.25 \text{ A right}}$, and $\boxed{I_5 = 1.25 \text{ A down}}$.

(b) $P = I^2 R = (1.25 \text{ A})^2 (8.0 \ \Omega) = \boxed{13 \text{ W}}$.

65. It takes $\boxed{\text{shorter}}$ than one time constant because the time constant is defined as the times it takes to charge the capacitor to 63% of its maximum charge.

69. (a) You should $\boxed{\text{(1) increase the capacitance}}$ to increase the time constant because $\tau = RC$.

(b) From $\tau = RC$, we have $R = \dfrac{\tau}{C} = \dfrac{2.0 \text{ s}}{1.0 \times 10^{-6} \text{ F}} = 2.0 \times 10^6 \ \Omega = \boxed{2.0 \text{ M}\Omega}$.

72. (a) The potential difference on the capacitor is zero immediately after the switch is closed because the capacitor is uncharged.

$V_C = V_0 \left(1 - e^{-t/\tau}\right) = V_0 \left(1 - e^0\right) = 0$. So the potential difference across the resistor is $\boxed{24 \text{ V}}$.

(b) $\boxed{0}$.

(c) $I = \dfrac{\varepsilon}{R} = \dfrac{24 \text{ V}}{6.0 \ \Omega} = \boxed{4.0 \text{ A}}$.

79. (a) An ammeter has very low resistance, so if it were connected in parallel in a circuit, the circuit current would be very high and the galvanometer could burn out.

(b) A voltmeter has very high resistance, so if it were connected in series in a circuit, it would read the voltage of the source because it has the highest resistance (most probably) and therefore the most voltage drop among the circuit elements.

82. (a) You should use $\boxed{\text{(1) a shunt resistor}}$. Since a galvanometer cannot allow a large current through it, current has to be diverted through another branch through the shunt resistor.

(b) From $I_g = \dfrac{IR_s}{r + R_s}$, we have

$$R_s = \frac{I_g r}{I - I_g} = \frac{(2000 \times 10^{-6}\,\text{A})(100\,\Omega)}{30\,\text{A} - 2000 \times 10^{-6}\,\text{A}} = 0.00667\,\Omega = \boxed{6.7\,\text{m}\Omega}.$$

88. (a) The internal resistance of the voltmeter should be $\boxed{\text{(3) infinite}}$.

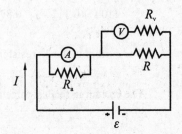

A voltmeter is connected in parallel with a circuit element. If its resistance is infinite, there will be no current through it; therefore, it does not affect the voltage across the circuit element it is measuring.

(b) The current reading I is the total current through the parallel combination R_p (R and R_v) so V/I gives the resistance of the parallel combination.

(c) The voltage reading is $V = IR_p = I\,\dfrac{R_v R}{R_v + R}$.

Therefore $R = \dfrac{V}{I - (V/R_v)}$.

(d) An ideal voltmeter has R_v approaching ∞, then $R = \dfrac{V}{I}$, i.e., the measurement is "perfect."

94. A conductor has very low resistance. The resistance of the wire between the feet is very small; so the voltage between the feet is small. Therefore the current through the bird is also.

96. The case is grounded so the potential of the case is always zero even if the hot wire accidentally touches the case.

100. (a) Three R's are in series, $\quad R_{s1} = 3R$.

This R_{s1} and R are in parallel, $\quad R_{p1} = \dfrac{(3R)R}{3R + R} = \dfrac{3}{4}R$.

This R_{p1} and two R are in series, $\quad R_{s2} = 2R + \dfrac{3}{4}R = \dfrac{11}{4}R$.

This R_{s2} and R are in parallel, $\quad R_{p2} = \dfrac{\frac{11}{4}R\,R}{\frac{11}{4}R + R} = \dfrac{11}{15}R$.

Finally, this R_{p2} and two R are in series, $\quad R_{s3} = 2R + \dfrac{11}{15}R = \boxed{41R/15 = 2.73R}$.

(b) The current in the two resistors closest to points A and B is $I = \dfrac{12.0\ \text{V}}{41(10\ \Omega)/15} = \boxed{0.439\ \text{A}}$.

The voltage across the combination without the first two resistors is $\dfrac{11}{15} \times (10\ \Omega)(0.439\ \text{A}) = 3.219\ \text{V}$.

The current in the third resistor closest to to points A and B is $\dfrac{3.219\ \text{V}}{10\ \Omega} = \boxed{0.322\ \text{A}}$.

The current in the 4th and 5th resistors closest to to points A and B is then $0.439\ \text{A} - 0.322\ \text{A} = \boxed{0.117\ \text{A}}$.

The voltage across the last four resistors farthest away from points A and B is

$\dfrac{3}{4} \times (10\ \Omega)(0.117\ \text{A}) = 0.8775\ \text{V}$.

The current in the 4th resistor farthest away from points A and B is $\dfrac{0.8775\ \text{V}}{10\ \Omega} = \boxed{0.0878\ \text{A}}$.

The current in the last three resistors farthest away from points A and B is $\dfrac{0.8775\ \text{V}}{3 \times 10\ \Omega} = \boxed{0.0293\ \text{A}}$.

V. Practice Quiz

1. Three resistors of 4.0 Ω, 6.0 Ω, and 10.0 Ω are connected in series. What is their equivalent resistance?

    (a) 20 Ω   (b) 7.3 Ω   (c) 6.0 Ω   (d) 4.0 Ω   (e) 1.9 Ω

2. Three resistors having values of 4.0 Ω, 6.0 Ω, and 10.0 Ω are connected in parallel. If the circuit is connected in series to a battery of 12.0 V and a resistor of 2.0 Ω, what is the current through the 10-Ω resistor?

    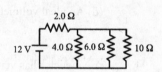

    (a) 0.59 A   (b) 1.0 A   (c) 2.7 A   (d) 11.2 A   (e) 16.0 A

3. The following three appliances are connected to a 120-V house circuit: (1) computer and printer, 350 W, (2) coffee pot, 650 W, and (3) microwave, 900 W. If all are operated at the same time what total current will they draw?

    (a) 0.063 A   (b) 2.9 A   (c) 5.4 A   (d) 7.5 A   (e) 16 A

4. Find the current in the 15-Ω resistor.

    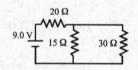

    (a) 0.10 A   (b) 0.13 A   (c) 0.20 A   (d) 0.26 A   (e) 0.30 A

5. What is the maximum number of 75-W lightbulbs you can connect in parallel in a 120-V home circuit without tripping the 15-A circuit breaker?

    (a) 15   (b) 18   (c) 21   (d) 24   (e) 27

6. What is the result from the Kirchhoff's junction rule for this figure?

(a) $I_2 = I_1 + I_3$

(b) $I_1 = I_2 + I_3$

(c) $I_3 = I_1 + I_2$

(d) $I_1 + I_2 + I_3 = 0$

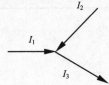

7. What is the current through the 2-Ω resistor in the circuit shown?

(a) 1 A   (b) 2 A   (c) 3 A   (d) 4 A   (e) 5 A

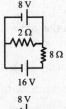

8. What is the current through the 8-Ω resistor in the circuit shown?

(a) 1 A   (b) 2 A   (c) 3 A   (d) 4 A   (e) 5 A

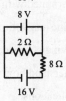

9. What is the equivalent resistance between points A and B of the resistors in the circuit?

(a) 0.443R   (b) 0.75R   (c) R   (d) 2.5R   (e) 4R

10. In the following simple circuit, G is a galvanometer and R is a resistor. What component of a circuit is this arrangement likely to be used to measure?

(a) voltage   (b) current   (c) resistance   (d) power   (e) energy

11. A voltage source of 10 V is connected to a series RC circuit with $R = 1.0$ MΩ and $C = 4.0$ μF. Initially, the capacitor is fully charged. Find the time required for the current in the circuit to decrease to 10% of its original value.

(a) 0.42 s   (b) 2.3 s   (c) 4.0 s   (d) 9.2 s   (e) 18 s.

12. A galvanometer with a full-scale reading of 600 μA and a coil resistance of 50 Ω is to be used to build a voltmeter designed to read 50 V at full scale. What is the required multiplier resistance?

(a) 0.6 mΩ   (b) 8.3 Ω   (c) 30 kΩ   (d) 83 kΩ   (e) 600 kΩ

Answers to Practice Quiz:

1.a 2.a 3.e 4.c 5.d 6.c 7.d 8.a 9.b 10.b 11.d 12.b

CHAPTER 19

Magnetism

I. Chapter Objectives

Upon completion of this chapter, you should be able to:

1. learn the force rule between magnetic poles, and explain how the direction of a magnetic field is determined with a compass.

2. define magnetic field strength, and determine the magnetic force exerted by a magnetic field on a moving charged particle.

3. understand how the magnetic force on charged particles is employed in various practical applications.

4. calculate the magnetic force on a current-carrying wire and the torque on a current-carrying loop, and explain the concept of a magnetic moment of a loop or coil.

5. explain the operation of various instruments whose functions depend on electromagnetic interactions between currents and magnetic fields.

6. understand the production of a magnetic field by electric currents, calculate the strength of the magnetic field in simple cases, and use the right-hand source rule to determine the direction of the magnetic field from the direction of the current that produces it.

7. explain how ferromagnetic materials enhance external magnetic fields, understand the concept of a material's magnetic permeability, explain how "permanent" magnets are produced, and how "permanent" magnetism can be destroyed.

*8. state some of the general characteristics of the Earth's magnetic field, explain some theories about its possible source, and discuss some of the ways in which the Earth's magnetic field affects our planet's local environment.

II. Chapter Summary and Discussion

1. Magnets, Magnetic Poles, and Magnetic Field Direction (Section 19.1)

Electricity and magnetism (electromagnetism) are manifestations of a single fundamental force or interaction, the electromagnetic force.

Magnets have two different poles or "centers" of force, which are designated as north and south poles. By the *law of poles*, like magnetic poles repel each other, and unlike magnetic poles attract each other. A magnet can create a **magnetic field**, similar to the electric field created by an electric charge.

The direction of a magnetic field $\vec{B}$ at any location is in the direction that the north pole of a compass at that location would point. Hence, the magnetic field lines outside a bar magnet are directed away from a magnetic north pole and toward a magnetic south pole.

2. Magnetic Field Strength and Magnetic Force (Section 19.2)

A **magnetic field** can exert forces only on *moving* charges. The *magnitude* of a magnetic field can be defined in terms of the magnetic force F on a moving charge (q) by $B = F/(qv \sin \theta)$, where v is the speed of the charge, and θ is the angle between the velocity vector and the magnetic field vector. The SI unit of magnetic field is the tesla (T). When $\vec{v}$ and $\vec{B}$ are perpendicular, the magnitude of the magnetic force is at its maximum, since when $\theta = 90°$, $F = qvB \sin \theta = qvB \sin 90° = qvB$ (maximum force).

The *direction* of the magnetic force on a charged particle is determined by the **right-hand force rule**: when the fingers of the right hand are pointed in the direction of $\vec{v}$ and then curled toward $\vec{B}$, the extended thumb points in the direction of $\vec{F}$ on a positive charge. (For a negative charge, the force is opposite this direction.).

Note: The magnetic force is always perpendicular to both the velocity vector and the magnetic field vector, or equivalently the magnetic force is perpendicular to a plane formed by the velocity and magnetic field vectors.

Example 19.1 An electron moves with a speed of 4.0×10^6 m/s along the $+x$-axis. It enters a region where there is a uniform magnetic field of 2.5 T, directed at an angle of 60° to the x-axis and lying in the x-y plane. Calculate the initial force and acceleration of the electron.

Solution: Given: $q = 1.60 \times 10^{-19}$ C, $m = 9.11 \times 10^{-31}$ kg, $v = 4.0 \times 10^6$ m/s, $B = 2.5$ T, $\theta = 60°$.

Find: $\vec{F}$ and $\vec{a}$.

According to the right-hand force rule, the direction of the magnetic force in the diagram is into the page ($-z$) for the electron because it has negative charge.

$F = qvB \sin \theta = (1.60 \times 10^{-19} \text{ C})(4.0 \times 10^6 \text{ m/s})(2.5 \text{ T}) \sin 60° = 1.4 \times 10^{-12}$ N.

By Newton's second law, $a = \dfrac{F}{m} = \dfrac{1.4 \times 10^{-12} \text{ N}}{9.11 \times 10^{-31} \text{ kg}} = 1.5 \times 10^{18}$ m/s^2.

The directions of both the magnetic force and acceleration are initially along the $-z$-axis.

Example 19.2 A proton has a speed of 4.5×10^6 m/s in a direction perpendicular to a uniform magnetic field, and the proton moves in a circle of radius 0.20 m. What is the magnitude of the magnetic field?

Solution: Given: $q = 1.60 \times 10^{-19}$ C, $m = 1.67 \times 10^{-27}$ kg, $v = 4.5 \times 10^6$ m/s, $r = 0.20$ m.

Find: B.

Because the magnetic force is always perpendicular to the velocity of the proton (right-hand force rule), it is the centripetal force that causes the particle to move in a circular path. That is, the centripetal force is provided by the magnetic force. By Newton's second law: $F_c = ma_c = mv^2/r = qvB$.

So $B = \dfrac{mv}{qr} = \dfrac{(1.67 \times 10^{-27} \text{ kg})(4.5 \times 10^6 \text{ m/s})}{(1.60 \times 10^{-19} \text{ C})(0.20 \text{ m})} = 0.23$ T.

3. Applications: Charged Particles in Magnetic Fields (Section 19.3)

Applications of charged particles in magnetic fields include the cathode ray tube (CRT), the mass spectrometer, and magnetohydrodynamics. A CRT is a vacuum tube that is used in oscilloscopes, computer monitors, and televisions. A mass spectrometer is a device used to separate isotopes, or atoms of different masses. It uses the different radii of circular orbits in a magnetic field by particles of the same charge but different mass.

In a mass spectrometer, a *velocity selector* is used to allow charged particles of only a certain velocity to enter the spectrometer. If a beam of charged particles is not deflected in the presence of a uniform electric field and a magnetic field mutually perpendicular to it, it must satisfy $v = \dfrac{E}{B_1} = \dfrac{V}{B_1 d}$, where B_1 is the magnetic field in the velocity selector, V is the voltage across it, and d is the separation of the plates in the velocity selector. The mass of the particle m depends on the radius of the circular orbit R in another magnetic field B_2: $m = \dfrac{qdB_1B_2}{V}R$.

Example 19.3 In a mass spectrometer, a singly charged particle has a speed of 1.0×10^6 m/s and enters a uniform magnetic field of 0.20 T. The radius of the circular orbit is 0.020 m.
(a) What is the mass of the particle?
(b) What is the kinetic energy of the particle?

Solution: Given: $q = 1.6 \times 10^{-19}$ C, $v = 1.0 \times 10^6$ m/s, $B_2 = 0.20$ T, $R = 0.020$ m.

Find: (a) m (b) K.

(a) Because $v = \dfrac{E}{B_1} = \dfrac{V}{B_1 d}$,

$$m = \frac{q d B_1 B_2}{V} R = \frac{q B_2 R}{V/(B_1 d)} = \frac{q B_2 R}{v} = \frac{(1.6 \times 10^{-19} \text{ C})(0.20 \text{ T})(0.020 \text{ m})}{1.0 \times 10^6 \text{ m/s}} = 6.4 \times 10^{-27} \text{ kg}.$$

The mass of a proton or a neutron is 1.67×10^{-27} kg and 6.4×10^{-27} kg $= 4 \times 1.67 \times 10^{-27}$ kg. Therefore this particle is likely the helium nucleus, which has two protons and two neutrons.

(b) $K = \frac{1}{2} m v^2 = \frac{1}{2}(6.4 \times 10^{-28} \text{ kg})(1.0 \times 10^6 \text{ m/s})^2 = 3.2 \times 10^{-16} \text{ J} = (3.2 \times 10^{-16} \text{ J}) \times \dfrac{1 \text{ eV}}{1.60 \times 10^{-19} \text{ J}}$

$\qquad = (2.0 \text{ keV}).$

4.   Magnetic Forces on Current-Carrying Wires (Section 19.4)

A magnetic field can exert a force F on a current-carrying wire. The magnitude of the magnetic force is given by $F = ILB \sin \theta$, where L is the length of the wire, I is the current in the wire, and θ is the angle between the current direction and the magnetic field vector. The direction of the force is determined by the right-hand force rule: when the fingers of the right hand are pointed in the direction of the conventional current and then turned or curled toward the vector $\vec{B}$, the extended thumb points in the direction of $\vec{F}$.

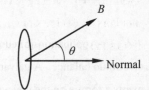

A magnetic field can exert torque on a current-carrying loop. The magnitude of the torque is equal to $\tau = NIAB \sin \theta$, where N is the number of turns (loops), I is the current in the loop, A is the area enclosed by the loop, and θ is the angle between the normal to the loop and the magnetic field vector. The quantity IA is often referred to as the **magnetic moment**, $m = IA$. The magnitude of the torque can then be expressed as $\tau = NmB \sin \theta$.

Note: The angle θ in the torque equation is between the *normal to the plane of the loop* and the magnetic field, *not* between the plane of the loop and the magnetic field.

Example 19.4   A wire carries a current of 6.0 A in a direction of 60° with respect to the direction of a magnetic field of 0.75 T. Find the magnitude of the magnetic force on a 0.50-m-length of the wire.

Solution: Given: $I = 6.0$ A, $\theta = 60°$, $B = 0.75$ T, $L = 0.50$ m.

Find: F.

$F = ILB \sin \theta = (6.0 \text{ A})(0.50 \text{ m})(0.75 \text{ T}) \sin 60° = 1.9$ N.

Example 19.5   A circular loop of wire of radius 0.50 m is in a uniform magnetic field of 0.30 T. The current in the loop is 2.0 A. Find the magnitude of the torque when

(a) the plane of the loop is parallel to the magnetic field.

(b) the plane of the loop is perpendicular to the magnetic field.

(c) the plane of the loop is at 30° to the magnetic field.

Solution: Given: $N = 1$, $r = 0.50$ m, $B = 0.30$ T, $I = 2.0$ A,

(a) $\theta = 90°$,   (b) $\theta = 0°$,   (c) $\theta = 90° - 30° = 60°$.

Find: τ in (a), (b), and (c).

The angle in the torque equation is between the normal to the plane of the loop and the magnetic field, not between the plane of the loop and the magnetic field.

(a) $\tau = NIAB \sin \theta = NA\pi r^2 B \sin \theta = (1)(2.0 \text{ A})(\pi)(0.50 \text{ m})^2(0.30 \text{ T}) \sin 90° = 0.47$ m·N.

(b) $\tau = (1)(2.0 \text{ A})(\pi)(0.50 \text{ m})^2(0.30 \text{ T}) \sin 0° = 0$ m·N.

(c) $\tau = (1)(2.0 \text{ A})(\pi)(0.50 \text{ m})^2(0.30 \text{ T}) \sin 60° = 0.41$ m·N.

5.   Applications: Current-Carrying Wires in Magnetic Fields (Section 19.5)

Applications of current-carrying wires in magnetic fields include the galvanometer, the dc motor, and the electronic balance. A galvanometer consists of a small coil. When there is a current in the coil, the magnetic field exerts a torque on the coil and therefore deflects the coil. The deflection of the coil is directly proportional to the current through the galvanometer. A dc motor is a device that converts electrical energy into mechanical energy. In an electronic balance, the magnetic force on a current-carrying wire replaces the known force in a conventional balance.

6.   Electromagnetism—The Source of Magnetic Fields (Section 19.6)

Generally, magnetic fields are produced by electric currents (moving charges). The magnitude of the magnetic field near *a long straight current-carrying wire* is given by $B = \dfrac{\mu_0 I}{2\pi d}$, where $\mu_0 = 4\pi \times 10^{-7}$ T·m/A is a constant called the *magnetic permeability of free space*, I is the current in the wire, and d is the perpendicular distance from the wire. The direction of the magnetic field is determined by the **right-hand source rule**: if a current-carrying wire is grasped with the right hand with the extended thumb pointing in the direction of the current (I), the curled fingers indicate the circular sense of the magnetic field.

The magnitude of the magnetic field at the *center of a circular current-carrying wire coil* is given by $B = \dfrac{\mu_0 NI}{2r}$, where r is the radius of the loop, and N is the number of turns (loops) in the coil. The direction of the magnetic field is determined by the right-hand source rule and is perpendicular to the plane of the loop at its center.

The magnitude of the magnetic field near the center of *a current-carrying solenoid* is given by $B = \dfrac{\mu_0 NI}{L}$, where N is the number of turns (loops) in the solenoid, and L is the length of the solenoid. The direction of the magnetic field is determined by the right-hand source rule as applied to one of the loops of the solenoid. The quantity $n = N/L$ is called the *linear turn density* (number of turns per unit length). The magnitude of the magnetic field can also be expressed in terms of n as $B = \mu_0 nI$. The magnetic field near the center of a solenoid is independent of the radius of the solenoid and is approximately constant if you stay away from the ends.

Note: Because magnetic field is a vector, you must use vector addition to find the net field if there are contributions from two or more sources.

Example 19.6 What current is required for a long straight wire to produce a magnetic field of magnitude equal to the strength of the Earth's magnetic field of about 5.0×10^{-5} T at a location 2.5 cm from the wire?

Solution: Given: $B = 5.0 \times 10^{-5}$ T, $d = 2.5$ cm $= 0.25$ m.

Find: I.

From $B = \dfrac{\mu_0 I}{2\pi d}$, we have $I = \dfrac{2\pi Bd}{\mu_0} = \dfrac{2\pi (5.0 \times 10^{-5} \text{ T})(0.025 \text{ m})}{4\pi \times 10^{-7} \text{ T·m/A}} = 6.3$ A.

Example 19.7 Calculate the magnitude of the magnetic field at the center of a 0.10 m-long solenoid that has 100 turns, and carries a current of 2.0 A.

Solution: Given: $N = 100$, $L = 0.10$ m, $I = 2.0$ A.

Find: B.

$B = \dfrac{\mu_0 NI}{L} = \dfrac{(4\pi \times 10^{-7} \text{ T·m/A})(100)(2.0 \text{ A})}{0.10 \text{ m}} = 2.5 \times 10^{-3}$ T $= 2.5$ mT.

Example 19.8   Two long parallel wires carry currents of 20 A and 5.0 A in opposite directions. The wires are separated by 0.20 m.

(a) What is the magnetic field midway between the two wires?

(b) At what point between the wires are the magnetic fields from the two wires the same?

Solution: Given: $I_1 = 20$ A, $I_2 = 5.0$ A, $d = 0.20$ m.

Find: (a) B at $d_1 = d_2 = 0.10$ m   (b) x where $B_1 = B_2$.

(a) According to the right-hand source rule, the magnetic fields between the wires due to the two wires are in the same direction (both are into the page according to the figure shown on the right). At midway between the two wires, the distance from either wire is 0.10 m.

$$B_1 = \frac{\mu_o I_1}{2\pi d_1} = \frac{(4\pi \times 10^{-7} \text{ T·m/A})(20 \text{ A})}{2\pi(0.10 \text{ m})} = 4.0 \times 10^{-5} \text{ T, and}$$

$$B_2 = \frac{(4\pi \times 10^{-7} \text{ T·m/A})(5.0 \text{ A})}{2\pi(0.10 \text{ m})} = 1.0 \times 10^{-5} \text{ T.}$$

Thus, the net magnetic field is $B = B_1 + B_2 = 4.0 \times 10^{-5}$ T $+ 1.0 \times 10^{-5}$ T $= 5.0 \times 10^{-5}$ T, into the page.

(b) Assume the contributions from both wires are the same at a distance $d_1 = x$ from the 20-A wire. Then, the distance from the 5.0-A wire is $d_2 = (0.20 \text{ m} - x)$.

From $B_1 = B_2$, we have $\dfrac{\mu_o I_1}{2\pi x} = \dfrac{\mu_o I_2}{(0.20 \text{ m} - x)}$.

So $I_1(0.20 \text{ m} - x) = I_2 x$, or $(20 \text{ A})(0.20 \text{ m} - x) = (5.0 \text{ A}) x.$

Solving for $x = 0.16$ m from the 20-A wire.

Forces exist between two parallel current-carrying wires, because the magnetic field produced by the current in one wire exerts a force on the other wire. The force per unit length can be calculated from

$\dfrac{F}{L} = \dfrac{\mu_o I_1 I_2}{2\pi d}$, where d is the distance between the parallel wires, and I_1 and I_2 are the currents in the wires, respectively. According to the right-hand rules (source plus force), the forces are attractive if the currents are in the same direction and repulsive if the currents are in opposite directions. (Can you show why?)

Integrated Example 19.9

Two long straight wires separated by a distance of 0.30 m carry currents in the same direction. (a) The magnetic force between the two wires is (1) attractive, (2) repulsive, or (3) along the wires. Explain. (b) If the current in one wire is 10 A, and the current in the other is 8.0 A, find the magnitude magnetic forces per unit length (per meter) between the two wires. (c) What if the currents are in opposite directions?

(a) Conceptual Reasoning:

Refer to the sketch on the right. According to the right-hand source rule, the magnetic field at wire 2 due to wire 1 is upward. Then according to the right-hand force rule, the magnetic force on wire 2 is toward wire 1. From Newton's third law, the force on wire 1 is therefore toward wire 2 so the force between the two wires is (1) attractive.

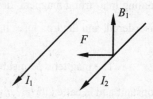

Another way to look at this situation is to focus on the area between the two wires and use the pole-force law. According to the right-hand source rule, the magnetic fields between the two wires by the two wires are opposite to each other. The field by wire 1 is upward and the field by wire 2 is downward. Therefore the two wires will attract each other.

(b) Quantitative Reasoning and Solution:

Given: $I_1 = 10$ A, $I_2 = 8.0$ A, $d = 0.30$ m.

Find: F/L.

The magnetic field at wire 2 by wire 1 is $B_2 = \dfrac{\mu_0 I_1}{2\pi d}$, so the magnitude of the force on wire 2 is equal to

$$F_2 = I_2 L B_2 = \frac{\mu_0 I_1 I_2}{2\pi d} L.$$

Therefore, the force per unit length is

$$\frac{F_2}{L} = \frac{\mu_0 I_1 I_2}{2\pi d} = \frac{(4\pi \times 10^{-7} \text{ T·m/A})(10 \text{ A})(8.0 \text{ A})}{2\pi (0.30 \text{ m})} = 5.3 \times 10^{-5} \text{ N/m}.$$

(c) If the currents are opposite, the magnetic force between the two wires is repulsive with the same magnitude as calculated in (b).

---

7. Magnetic Materials (Section 19.7)

Ferromagnetic materials are those in which the electron spins "lock" together to create **magnetic domains** where the magnetic fields of individual electrons add constructively. These materials are easily magnetized. The magnetic permeability of these materials is many times that of free space, reflecting the fact that they enhance an external magnetic field. In an external magnetic field, the domains parallel to the field grow at the expense of other domains, and the orientation of some domains may become more aligned with the field.

Common ferromagnetic materials are iron, nickel, and cobalt. Iron is commonly used in the cores of electromagnets. This type of iron is termed "soft" iron because the domains become unaligned and the iron unmagnetized when the external field is removed. "Hard" iron retains some magnetism after a field is removed, and this type of iron is used to make "permanent" magnets by heating the iron above the **Curie temperature** and cooling in a strong magnetic field. Above the Curie temperature, the magnetic domains become thermally disordered, and the iron loses its "permanent" magnetic field.

The magnetic permeability is defined as $\mu = \kappa_m \mu_o$, where κ_m is the magnetic analog of the dielectric constant and is called the *relative permeability*. $\kappa_m = 1$ for free space (vacuum).

*8.   Geomagnetism: The Earth's Magnetic Field (Section 19.8)

The Earth's magnetic field resembles that which would be produced by a large interior bar magnet (with the magnet's south pole near the Earth's north geographic pole and the magnet's north pole near the Earth's south geographic pole; however, this is not possible because the Earth's interior temperature is above the Curie temperature of ferromagnetic materials. Scientists associate the Earth's magnetic field with motions (electric current) in its liquid outer core.

The magnetic north pole and the geographic south pole do not coincide (nor do the magnetic south pole and geographic north pole), and the magnetic poles "wander" or move about over periods of hundreds of thousands of years. The deviation between the north-south magnetic poles and the south-north geographic poles is called the *magnetic declination*. There is evidence that the Earth's magnetic poles have reversed periodically many times over long geologic time periods.

Charged particles from the Sun and other cosmic rays can be trapped in the Earth's magnetic field, and regions or concentrations of these charged particles are called *Van Allen belts*. It is believed that the recombination of ionized air molecules and electrons that have been ionized by particles from the lower belt give rise to the aurora borealis (northern lights) and aurora australis (southern lights).

III.   Mathematical Summary

Magnitude of the Magnetic Force on a Charged Particle	$F = qvB \sin \theta$ (19.3) θ — angle between $\vec{v}$ and $\vec{B}$	Computes the magnitude of the magnetic force on a moving charge.
Magnitude of Force on a Current-Carrying Wire	$F = ILB \sin \theta$ (19.7) θ — angle between $\vec{I}$ and $\vec{B}$	Computes the magnitude of the magnetic force on a straight current-carrying wire.
Magnitude of the Torque on a Current-Carrying Coil	$\tau = NIAB \sin \theta$ (19.9) θ — angle between $\vec{m}$ and $\vec{B}$	Computes the magnitude of the magnetic torque on a current-carrying coil of N loops.

Magnitude of the Magnetic Field by a Long, Straight Wire	$B = \dfrac{\mu_o I}{2\pi d}$ (19.12) $\mu_o = 4\pi \times 10^{-7}$ T·m/A	Calculates the magnitude of the magnetic field due to a long straight current-carrying wire.
Magnitude of the Magnetic Field at the Center of a Circular Loop	$B = \dfrac{\mu_o NI}{2r}$ (19.13)	Calculates the magnitude of the magnetic field at the center of N circular loops of current-carrying wire.
Magnitude of the Magnetic Field at the Center of a Solenoid	$B = \dfrac{\mu_o NI}{L}$ (19.14)	Computes the magnitude of the magnetic field along the axis of a long solenoid (along the axis).

IV. Solutions of Selected Exercises and Paired Exercises

4. The magnet would attract the unmagnetized iron bar when a pole end is placed at the center of its long side. If the end of the unmagnetized bar were placed at the center of the long side of the magnet, it would not be attracted.

14. (a) According to the right-hand force rule, the magnetic force is directed $\boxed{\text{(1) into the page}}$.

(b) From $F = qvB \sin \theta$, we have $B = \dfrac{F}{qv \sin \theta} = \dfrac{20 \text{ N}}{(0.25 \text{ C})(2.0 \times 10^2 \text{ m/s}) \sin 90°} = \boxed{0.40 \text{ T}}$.

18. (a) According to the right-hand force rule and the fact that electron carries negative charge, the magnetic field is directed in the $\boxed{\text{(4) } -z}$ direction.

(b) From $F = qvB \sin \theta$, we have

$B = \dfrac{F}{qv \sin \theta} = \dfrac{5.0 \times 10^{-19} \text{ N}}{(1.6 \times 10^{-19} \text{ C})(3.0 \times 10^6 \text{ m/s}) \sin 90°} = \boxed{1.0 \times 10^{-6} \text{ T}}$.

30. (a) $v = \dfrac{V}{Bd} = \dfrac{E}{B} = \dfrac{3000 \text{ N/C}}{0.030 \text{ T}} = \boxed{1.0 \times 10^5 \text{ m/s}}$.

(b) It is the $\boxed{\text{same speed}}$, 1.0×10^5 m/s, because it is $\boxed{\text{independent of charge}}$ on the particle.

32. $v = \dfrac{V}{Bd} = \dfrac{E}{B} = \dfrac{1.0 \times 10^3 \text{ V/m}}{0.10 \text{ T}} = 1.0 \times 10^4$ m/s. Here the magnetic force provides centripetal force for the

circular motion. Since $F = qvB \sin \theta = m \dfrac{v^2}{R}$,

$m = \dfrac{qBR \sin \theta}{v} = \dfrac{(1.6 \times 10^{-19} \text{ C})(0.10 \text{ T})(0.012 \text{ m}) \sin 90°}{1.0 \times 10^4 \text{ m/s}} = \boxed{1.9 \times 10^{-26} \text{ kg}}$.

38. They attract each other because the magnetic fields created by the two wires in a region between the wires are opposite. The magnitudes of the forces on each wire are the same by Newton's third law.

42. (a) The magnetic force is in the $(3) +y-$ direction according to the right-hand force rule (upward as $+z$).

   (b) $F = ILB \sin \theta = (5.0 \text{ A})(1.0 \text{ m})(0.30 \text{ T}) \sin 90° = \boxed{1.5 \text{ N}}$.

44. (a) To the $\boxed{\text{right}}$.　　　　　　(b) $\boxed{\text{Toward top of the page}}$.

   (c) $\boxed{\text{Into the page}}$.　　　　　(d) To the $\boxed{\text{left}}$.

   (e) $\boxed{\text{Into or out of the page}}$.

50. $\dfrac{F}{L} = \dfrac{\mu_0 I_1 I_2}{2\pi d} = \dfrac{(4\pi \times 10^{-7} \text{ T·m/A})(15 \text{ A})^2}{2\pi (0.15 \text{ m})} = \boxed{3.0 \times 10^{-4} \text{ N/m, repulsive}}$. It is repulsive because the

   currents are opposite and so the fields in between the wires are in the same direction.

51. (a) The forces on the wires are $\boxed{(1) \text{ attractive}}$. Assume the currents are upward in both wires. The

   magnetic field by the left wire on the right wire is directed into the page, according to the right-hand source rule. Then the magnetic force on the right wire is to the left according to the right-hand force rule. Vice versa, the force on the left wire is to the right so they attract.

   (b) $\dfrac{F}{L} = \dfrac{\mu_0 I_1 I_2}{2\pi d} = \dfrac{(4\pi \times 10^{-7} \text{ T·m/A})(2.0 \text{ A})(4.0 \text{ A})}{2\pi (0.24 \text{ m})} = \boxed{6.7 \times 10^{-6} \text{ N/m}}$.

56. (a) Use the result of Exercise 51(a). The top wire must attract the lower wire so it can stay in equilibrium. For the forces to attract, the currents should be in $\boxed{(1) \text{ the same}}$ direction.

   (b) Magnetic force cancels gravity. $F = mg$. So $\dfrac{F}{L} = \dfrac{\mu_0 I_1 I_2}{2\pi d} = \dfrac{mg}{L}$.

   For a 1.0-m length, $I = I_1 = I_2 = \sqrt{\dfrac{2\pi d mg}{\mu_0}} = \sqrt{\dfrac{2\pi (0.020 \text{ m})(1.5 \times 10^{-3} \text{ kg})(9.80 \text{ m/s}^2)}{4\pi \times 10^{-7} \text{ T·m/A}}} = \boxed{38 \text{ A}}$.

66. Not necessarily. The magnetic field in a solenoid depends on the current in it and the turn per unit length, not just the number of turns. For example, if the 200 turns is over 0.20 m and the 100 turns is over 0.10 m, then they will have the same turns per unit length and then the same magnetic field.

70. $B = \dfrac{\mu_0 I}{2\pi d} = \dfrac{(4\pi \times 10^{-7} \text{ T·m/A})(2.5 \text{ A})}{2\pi (0.25 \text{ m})} = \boxed{2.0 \times 10^{-6} \text{ T}}$.

75. The magnitudes at both locations are the same.

Calculate for the location to the right of I_2. $B = \dfrac{\mu_0 I}{2\pi d}$.

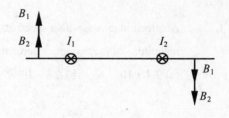

$B_1 = \dfrac{(4\pi \times 10^{-7} \text{ T·m/A})(1.5 \text{ A})}{2\pi (0.35 \text{ m})} = 8.57 \times 10^{-7} \text{ T},$

$B_2 = \dfrac{(4\pi \times 10^{-7} \text{ T·m/A})(1.5 \text{ A})}{2\pi (0.15 \text{ m})} = 2.0 \times 10^{-6} \text{ T}.$

So the net magnetic field is $B = B_1 + B_2 = 7.57 \times 10^{-7} \text{ T} + 2.0 \times 10^{-6} \text{ T} = \boxed{2.9 \times 10^{-6} \text{ T}}$.

78. The Earth's magnetic field at the equator is about $0.40 \text{ G} = 4.0 \times 10^{-5} \text{ T}$.

Since $B = \dfrac{\mu_0 I}{2r}$, $I = \dfrac{2rB}{\mu_0} = \dfrac{2(0.10 \text{ m})(4.0 \times 10^{-5} \text{ T})}{4\pi \times 10^{-7} \text{ T·m/A}} = \boxed{6.4 \text{ A}}$.

84. The two magnetic fields by the two wires are perpendicular to each other.

$B_1 = \dfrac{\mu_0 I_1}{2\pi d_1} = \dfrac{(4\pi \times 10^{-7} \text{ T·m/A})(15 \text{ A})}{2\pi (0.10 \text{ m})} = 3.0 \times 10^{-5} \text{ T}$ along the page out.

$B_2 = B_1 = 3.0 \times 10^{-5} \text{ T}$ into the page.

So the net magnetic field is

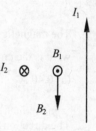

$B = \sqrt{B_1^2 + B_2^2} = \sqrt{(3.0 \times 10^{-5} \text{ T})^2 + (3.0 \times 10^{-5} \text{ T})^2} = \boxed{4.2 \times 10^{-5} \text{ T}}$.

92. (a) $B = \mu n I = K_m \mu_0 n I = (2000)(4\pi \times 10^{-7} \text{ T·m/A})(100 \times 10^2 /\text{m})(0.040 \text{ A}) = \boxed{1.0 \text{ T}}$.

(b) From $B_0 = \mu_0 n I$, we have $\dfrac{B}{B_0} = K_m = 2.0 \times 10^3$. So $\boxed{B = (2.0 \times 10^3)B_0}$.

98. (a) $\boxed{\text{Zero}}$. (b) $\boxed{\text{Up}}$. (c) $\boxed{\text{East}}$.

102. (a) The force is directed $\boxed{(1) \text{ west}}$ according to the right-hand force rule (electron carries negative charge).

(b) $F = qvB \sin \theta = (1.6 \times 10^{-19} \text{ C})(10^3 \text{ m/s})(5.0 \times 10^{-5} \text{ T}) \sin 90° = \boxed{8.0 \times 10^{-21} \text{ N}}$.

108. (a) The magnetic field could be zero $\boxed{(4) \text{ inside the smaller one or outside the larger one}}$.

(b) $B_1 = N_1 \dfrac{\mu_0 I_1}{2r_1} = B_2 = N_2 \dfrac{\mu_0 I_2}{2r_2}$.

V. Practice Quiz

1. An electron moves with a speed of 3.0×10^4 m/s in a uniform magnetic field of 0.20 T. What is the magnitude of the maximum force it can experience?
   (a) 2.4×10^{-16} N (b) 4.8×10^{-16} N (c) 9.6×10^{-16} N (d) 1.9×10^{-15} N (e) zero

2. A wire of length 2.0 m carrying a current of 0.60 A is oriented at a 35° angle to a uniform magnetic field of 0.50 T. What is the magnitude of the force it experiences?
   (a) zero (b) 0.30 N (c) 0.34 N (d) 0.49 N (e) 0.60 N

3. A proton moving along the +x-axis enters a region where there is a uniform magnetic field in the +y-direction. What is the direction of the magnetic force on the proton, if the +x-axis is to the right, the +y-axis points upward, and the +z-axis is out of the page?
   (a) +z-direction (b) −z-direction (c) +y-direction (d) −y-direction (e) −x-direction

4. The magnetic field at the center of a solenoid is 25 mT. If the solenoid has 500 turns and is 0.10 m long, what is the current through it?
   (a) 0.40 A (b) 4.0 A (c) 0.40 mA (d) 5.0 mA (e) 8.0 mA

5. The direction of the force on a current-carrying wire in a magnetic field is
   (a) perpendicular to the current. (b) perpendicular to the magnetic field. (c) parallel to the current.
   (d) both (a) and (b). (e) neither (a) nor (b).

6. A circular loop carrying a current of 2.0 A is in a magnetic field of 0.35 T. The loop has an area of 1.2 m^2, and its plane is oriented at a 53° angle to the field. What is the magnitude of the magnetic torque on the loop?
   (a) 46 m·N (b) 0.67 m·N (c) 0.51 m·N (d) 0.10 m·N (e) zero

7. Two long parallel wires carry equal currents. The magnitude of the force between the wires is F. The current in each wire is then doubled while the distance between them is halved. What is the magnitude of the new force between the two wires?
   (a) F (b) $2F$ (c) $4F$ (d) $8F$ (e) $16F$

8. A proton travels from rest through a voltage of 1.0 kV and then moves into a magnetic field of 0.040 T perpendicular to its velocity. What is the radius of the proton's resulting orbit?
   (a) 0.080 m (b) 0.11 m (c) 0.14 m (d) 0.17 m (e) 0.21 m

9.   A thin metal rod 1.0 m long has a mass of 0.050 kg and is in a magnetic field of 0.10 T. What minimum current in the rod is needed in order for the magnetic force to cancel the weight of the rod?

(a) 0.58 A   (b) 1.2 A   (c) 2.5 A   (d) 4.9 A   (e) 9.8 A

10.   What is the magnitude of the magnetic force on a stationary electron in a uniform magnetic field of 0.20 T?

(a) 1.6×10^{-20} N   (b) 1.6×10^{-21} N   (c) 3.2×10^{-20} N   (d) 3.2×10^{-21} N   (e) zero

11.   A coil of ten circular loops of radius 5.0 cm carries a current of 2.5 A clockwise, as viewed from above the plane of the coil. What is the magnitude of the magnetic field at the center of the coil?

(a) 6.3×10^{-4} T   (b) 3.1×10^{-4} T   (c) 6.3×10^{-5} T   (d) 3.1×10^{-5} T   (e) zero

12.   Two long parallel wires, separated by 12 cm, carry currents of 8.0 A and 2.0 A in opposite directions. What is the magnitude of the magnetic field midway between the wires?

(a) 3.3×10^{-5} T   (b) 2.0×10^{-5} T   (c) 1.3×10^{-5} T   (d) 2.7×10^{-5} T   (e) 6.7×10^{-6} T

Answers to Practice Quiz:

1. c 2. c 3. a 4. b 5. d 6. c 7. d 8. b 9. d 10. e 11. b 12. a

CHAPTER 20

Electromagnetic Induction and Waves

I.   Chapter Objectives

Upon completion of this chapter, you should be able to:

1.   define magnetic flux and explain how an induced emf is created, and determine induced emfs and currents.

2.   understand the operation of electrical generators and calculate the emf produced by an ac generator, and explain the origin of back emf and its effect on the behavior of motors.

3.   explain transformer action in terms of Faraday's law, calculate the output of step-up and step-down transformers, and understand the importance of transformers in electric energy delivery systems.

4.   explain the physical nature, origin, and propagation of electromagnetic waves, and describe some of the properties and uses of various types of electromagnetic waves.

II.   Chapter Summary and Discussion

1.   Induced Emf: Faraday's Law and Lenz's Law (Section 20.1)

Magnetic flux is a measure of the number of magnetic field lines through a particular loop area:
$\Phi = BA \cos \theta$, where B is the magnetic field, A is the area of the loop, and θ is the angle between the magnetic field vector and the normal to the loop area. The SI unit of magnetic flux is T·m^2.

Note: The angle θ is the angle between the magnetic field and the *normal to the plane of the loop*, *not* between the magnetic field and the plane of the loop.

Electromagnetic induction refers to the creation of induced emf's whenever the magnetic flux through a loop or coil changes. The magnitude of induced emf in a conducting coil depends on the time rate change of magnetic flux through the loop, $\varepsilon = -N \dfrac{\Delta \Phi}{\Delta t}$, where N is the number of loops. This relationship is known as

Faraday's law of induction. Because $\Phi = BA \cos \theta$, any change in B, A, or θ will result in an induced emf.

Faraday's law can be written in detail as $\varepsilon = -N \left[\dfrac{\Delta B}{\Delta t} A \cos \theta + B \dfrac{\Delta A}{\Delta t} \cos \theta + BA \dfrac{\Delta(\cos \theta)}{\Delta t} \right]$. There are three different ways to produce induced emf.

When a conductor of length L moves perpendicular to a magnetic field B with a speed v, the magnitude of the induced emf is called *motional emf* and is given by $\varepsilon = BLv$.

The minus sign in Faraday's law of induction gives the polarity of the induced emf, which is found by **Lenz's law**: when a change in magnetic flux induces an emf in a conducting loop, the induced emf produces a current in such a direction as to create a magnetic field that tends to oppose the change in flux. That is, an induced emf gives rise to a current whose magnetic field opposes the change in magnetic flux that produced it.

Example 20.1   A circular loop of radius 0.20 m is rotating in a uniform magnetic field of 0.20 T. Find the magnetic flux through the loop when the plane of the loop and the magnetic field vector

(a) are parallel, (b) are perpendicular, (c) are at 60° to each other.

Solution: Given: $r = 0.10$ m, $B = 0.20$ T,   (a) $\theta = 90°$,   (b) $\theta = 0°$,   (c) $\theta = 90° - 60° = 30°$.

Find: Φ in (a), (b), and (c).

The angle θ in the magnetic flux definition is the angle between the magnetic field and the *normal* to the plane of the loop, *not* between the magnetic field and the plane of the loop.

(a) $\Phi = BA \cos \theta = B \pi r^2 \cos \theta = (0.20 \text{ T})(\pi)(0.20 \text{ m})^2 \cos 90° = 0.$

(b) $\Phi = (0.20 \text{ T})(\pi)(0.20 \text{ m})^2 \cos 0° = 2.5 \times 10^{-2}$ T·m^2.

(c) $\Phi = (0.20 \text{ T})(\pi)(0.20 \text{ m})^2 \cos 30° = 2.2 \times 10^{-2}$ T·m^2.

Example 20.2   A coil is wrapped with 100 turns of wire on a square frame with 18 cm sides. A magnetic field is applied perpendicular to the plane of the coil. If the field changes uniformly from 0 to 0.50 T in 8.0 s, find the average value of the magnitude of the induced emf.

Solution: Given: $N = 100$, $d = 0.18$ m, $T_i = 0$, $T_f = 0.50$ T, $\Delta t = 8.0$ s, $\theta = 0°$.

Find: ε.

From Faraday's law of induction, the magnitude of the induced emf is

$$\varepsilon = N \frac{\Delta B}{\Delta t} A \cos \theta = (100) \times \frac{0.50 \text{ T} - 0}{8.0 \text{ s}} \times (0.18 \text{ m})^2 \cos 0° = 0.21 \text{ V}.$$

Both $\dfrac{\Delta A}{\Delta t}$ and $\dfrac{\Delta(\cos \theta)}{\Delta t}$ are zero because A and θ do not change.

Example 20.3   A square coil of wire with 15 turns and an area of 0.40 m² is placed parallel to a magnetic field of 0.75 T. The coil is flipped so its plane is perpendicular to the magnetic field in 0.050 s. What is the magnitude of the average induced emf?

Solution: For B, A, and θ in this example, only θ changes.

Given: $N = 15$, $A = 0.40$ m², $\Delta t = 0.050$ s, $B = 0.75$ T, $\theta_i = 90°$, $\theta_f = 0°$.

Find: $\mathcal{E}$.

From Faraday's law of induction, the magnitude of the induced emf is

$$\mathcal{E} = NBA \frac{\Delta(\cos \theta)}{\Delta t} = (15)(0.75 \text{ T})(0.40 \text{ m}^2) \times \frac{\cos 0° - \cos 90°}{0.050 \text{ s}} = 90 \text{ V}.$$

Both $\dfrac{\Delta B}{\Delta t}$ and $\dfrac{\Delta A}{\Delta t}$ are zero because B and A do not change.

Example 20.4   An airplane with a wing span of 50 m flies horizontally with a speed of 200 m/s above the Earth at a location where the downward component of the Earth's magnetic field is 6.0×10^{-5} T. Find the magnitude of the induced emf between the wing's tips.

Solution: Given: $L = 50$m, $B = 6.0 \times 10^{-5}$ T, $v = 200$ m/s.

Find: $\mathcal{E}$.

The magnitude of the motional emf is $\mathcal{E} = BLv = (6.0 \times 10^{-5} \text{ T})(50 \text{ m})(200 \text{ m/s}) = 0.60 \text{ V}.$

Example 20.5   Find the direction of the induced current through R in the figures shown.

(a)   (b)

(c)   (d)

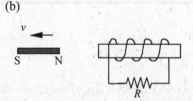

(e)   (f)

Magnetic field is into the page   Magnetic field is into the page

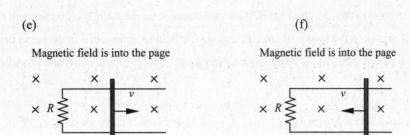

Solution:

We know from Lenz's law that the magnetic field produced by the induced current will tend to oppose or resist the change in flux.

(a) As the north pole of the magnet approaches the coil the field lines going into the coil increase to the right. The induced current in the coil produces a magnetic field that opposes the increase, so the left end of the coil will be a north pole. By the right-hand source rule, the induced current is from right to left through the resistor.

(b) As the north pole of the magnet moves away from the coil the field lines going into the coil decrease. The induced current in the coil produces a magnetic field that opposes the decrease, so the left end of the coil will be a south pole. By the right-hand source rule, the induced current is from right to left through the resistor.

Now, try to repeat parts (a) and (b) by flipping the magnet, that is, placing the south pole near the coil.

(c) When the switch is closed, the magnetic field due to the coil on the left increases, so the field lines going through the coil on the right increase. The induced current in the coil on the right will produce a magnetic field to oppose that increase, so the left end of the coil on the right will be a south pole. Therefore, the current is from left to right.

(d) When the switch is opened the magnetic field due to the coil on the left decreases, so the field lines going through the coil on the right decrease. The induced current in the coil on the right will produce a magnetic field to oppose that decrease, so the left end of the coil on the right will be a north pole. Therefore, the current is from right to left.

(e) As the rod moves to the right the area of the loop increases, so the magnetic flux through the area also increases. The induced current in the loop will produce a magnetic field to oppose that increase, so the magnetic field due to the induced current is out of the page. By the right-hand source rule, the current is from bottom to top through the rod.

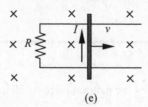
(e)

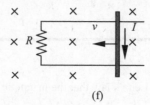
(f)

(f) As the rod moves to the left the area of the loop decreases, so the magnetic flux through the area also decreases into the page. The induced current in the loop will produce a magnetic field to oppose that decrease, so the magnetic field due to the induced current is into the page. The current is from top to bottom through the rod.

---

2.   Electric Generators and Back Emf (Section 20.2)

A **generator** is a device for converting mechanical energy into electrical energy. In principle, it is the reverse of an electrical motor. An **ac generator** or alternator produces **alternating current** (ac), meaning that the polarity of the voltage and direction of the current change periodically. A **dc generator** produces direct current (dc). When coils are rotated in a magnetic field (or the field is rotated while the coils are kept fixed) an emf is induced in the coils and can be used to create an electric current. This creation of emf and current is due to the term $\frac{\Delta(\cos \theta)}{\Delta t}$ in Faraday's law of induction.

The emf produced in an ac generator is expressed as $\varepsilon = \varepsilon_0 \sin \omega t = \varepsilon_0 \sin 2\pi f t$, where $\omega = 2\pi f$ is the angular frequency and $\varepsilon_0 = NBA\omega$ is the maximum emf (N is the number of turns of the coil, B is the magnetic field, A is the area of the coil).

A **back emf** is a reverse emf created by induction in a motor when its armature is rotated in a magnetic field. The direction is opposite the applied emf due to Lenz's law. The back emf partially cancels the voltage that drives the motor, so the current through the armature when the armature is rotating (operating) is smaller than when it is not (at start-up). When a motor is at operating speed, the back emf is given by $\varepsilon_b = V - IR$, where V is the applied driving voltage, I is the current through the armature, and R is the resistance of the armature.

Example 20.6   An ac generator consists of 200 turns of wire of area 0.090 m^2 and total resistance 12 Ω. The loop rotates in a magnetic field of 0.50 T at a constant angular speed of 60 revolutions per second. Find the maximum induced emf and the maximum current.

Solution: Given: $N = 200$, $A = 0.090$ m^2, $R = 12$ Ω, $B = 0.50$ T, $f = 60$ Hz.

Find: ε_o.

Because the generator is rotating at 60 rev/s, the frequency is 60 cycles/s = 60 Hz.

The angular frequency is $\omega = 2\pi f = 2\pi(60 \text{ Hz}) = 120\pi$ rad/s,

so the maximum emf is $\varepsilon_o = NAB\omega = (200)(0.090 \text{ m}^2)(0.50 \text{ T})(120\pi \text{ rad/s}) = 3.4 \times 10^3$ V = 3.4 kV.

Integrated Example 20.7

A dc motor of internal resistance 5.0 Ω is connected to a 24-V power supply and operates with an operating current of 2.0 A. (a) When the motor starts, its start-up current is (1) greater than, (2) the same as, or (3) less than the operating current when the motor is running continuously. Explain. (b) Calculate the start-up current. (c) What are the back emf when the motor is running at full speed and at half speed, respectively?

(a) Conceptual Reasoning:

(a) At start-up, there is no back emf in the motor because the motor is not rotating so the voltage across the motor is the same as the applied voltage. When the motor runs, it develops a back emf that is opposite to the applied voltage. Therefore, the true voltage across motor is less than the applied voltage. In other words, the voltage across the motor is higher at start-up. Therefore, the start-up current is (1) greater than the operating current.

(b) Quantitative Reasoning and Solution:

Given: $R = 5.0$ Ω, $V = 24$ V, $I = 2.0$ A.

Find: (b) I (start-up)   (c) ε_b at full speed and at half speed.

(b) At start-up, the voltage across the motor is equal to the applied voltage.

So the current is $I = \dfrac{V}{R} = \dfrac{24 \text{ V}}{5.0 \text{ Ω}} = 4.8$ A. As expected, this is greater than the 2.0-A operating current.

(c) At full operating speed, the back emf is equal to $\varepsilon_b = V - IR = 24$ V $- (2.0$ A$)(5.0$ Ω$) = 14$ V.

From Faraday's law or the generator equation, the induced emf (back emf is induced emf, after all) is directly proportional to the angular speed, so at half speed, the back emf is also halved, to (14 V)/2 = 7.0 V.

3. Transformers and Power Transmission (Section 20.3)

A **transformer** is a device that changes the ac voltage and current supplied to it by means of induction. (A transformer cannot work with dc voltage. Why?) It consists of two coils, a **primary coil** (the coil on the input side) and a **secondary coil** (the coil on the output side) and usually an iron core. If the voltage on the secondary is higher than that on the primary, it is a **step-up transformer**. A **step-down transformer** has lower voltage on the secondary side. Whether a transformer is a step-up or step-down transformer is solely determined by the relative number of turns in the primary and secondary coils. A step-up transformer has more turns in the secondary, whereas a step-down transformer has fewer turns in the secondary. According to the conservation of energy, if the voltage is increased, the current is reduced, and vice versa. A real (in contrast with an ideal one) transformer has I^2R energy losses (Joule heating) in the windings and energy loss caused by **eddy currents** in the transformer core. The number of turns N, voltage V, and current I ratios are related by $\dfrac{V_s}{V_p} = \dfrac{I_p}{I_s} = \dfrac{N_s}{N_p}$, where the subscript p stands for primary, and s for secondary.

In power transmission, the voltage is stepped up before transmission so as to decrease the current and the I^2R losses in the transmission lines (see Example 20.8, following). The voltage is then stepped down before the power is supplied to businesses and homes.

Example 20.8   A generator produces 60 A of current at 120 V. The voltage is stepped up to 4500 V by a transformer and transmitted through a power line of total resistance 1.0 Ω. Find:

(a) the number of turns in the secondary coil if the primary coil has 200 turns.

(b) the power lost in the transmission line.

(c) the power that would be lost in the transmission line if no transformer were used.

Solution: Given: $I_p = 60$ A, $V_p = 120$ V, $V_s = 4500$ V, $R = 1.0\ \Omega$,   (a) $N_p = 200$.

Find: (a) N_s (b) P_{loss} (c) P_{loss} (no transformer).

(a) From $\dfrac{V_s}{V_p} = \dfrac{I_p}{I_s} = \dfrac{N_s}{N_p}$, we have $N_s = \dfrac{V_s}{V_p} N_p = \dfrac{4500\ \text{V}}{120\ \text{V}} \times (200) = 7500$ turns.

(b) $I_s = \dfrac{V_p}{V_s} I_p = \dfrac{120\ \text{V}}{4500\ \text{V}} \times (60\ \text{A}) = 1.6$ A.

So the total power loss in the transmission lines is $P_{loss} = I^2 R = (1.6\ \text{A})^2 (1.0\ \Omega) = 2.6$ W.

The total power generated is $P_{total} = (60\ \text{A})(120\ \text{V}) = 7200$ W.

Therefore, the percentage loss is $\dfrac{2.6\ \text{W}}{7200\ \text{W}} = 0.036\%$.

(c) If no transformer were used, the current in transmission would be 60 A.

The total power loss would be $P_{loss} = (60 \text{ A})^2 (1.0 \ \Omega) = 3600 \text{ W}$.

So the percentage loss would be $\dfrac{3600 \text{ W}}{7200 \text{ W}} = 50\%$!

4. Electromagnetic Waves (Section 20.4)

An **electromagnetic wave** (or electromagnetic **radiation**) consists of mutually perpendicular, time-varying electric and magnetic fields that propagate at a constant speed in vacuum (the speed of light, $c = 3.00 \times 10^8$ m/s). **Maxwell's equations** are a set of four equations that describe all magnetic and electric field phenomena. Radiation carries energy and momentum and thus can exert force and therefore **radiation pressure** (radiation force per unit area).

The different types of electromagnetic radiation differ only in frequency f, and thus in wavelength λ, because they all travel at the same speed, c, in a vacuum, and $c = \lambda f$. The major types of electromagnetic radiation in order of increasing frequency or decreasing wavelength are power waves, radio and TV waves, microwaves, infrared radiation, visible light, ultraviolet radiation, X rays, and gamma rays.

Example 20.9   A Doppler radar indicates that a transmitted pulse is returned by clouds as an echo 40 μs after transmission. How far away are the clouds?

Solution: Given: $\Delta t_{total} = 40 \ \mu s = 40 \times 10^{-6}$ s, $c = 3.00 \times 10^8$ m/s.

           Find: d.

Because the time is for the echo (forward and back), the time for one-way distance is $\Delta t_{1/2} = 20 \times 10^{-6}$ s.

Therefore, the distance is $d = c \Delta t_{1/2} = (3.00 \times 10^8 \text{ m/s})(20 \times 10^{-6} \text{ s}) = 6.0 \times 10^3 \text{ m} = 6.0 \text{ km}$.

Example 20.10   The call number of a radio station indicates the frequency of the carrier wave. For the AM band, the frequency is expressed in kilohertz (kHz), and for the FM band, the frequency is in megahertz (MHz). What are the wavelengths for the carrier waves of the following two radio stations?
           (a) WBRN 1460 (AM)   (b) WCKC 107 (FM)

Solution: Given: $c = 3.00 \times 10^8$ m/s,   (a) $f = 1460$ kHz,   (b) $f = 107$ MHz.

           Find: λ in (a) and (b).

(a) From $c = \lambda f$, we have $\lambda = \dfrac{c}{f} = \dfrac{3.00 \times 10^8 \text{ m/s}}{1460 \times 10^3 \text{ Hz}} = 205$ m.

(b) $\lambda = \dfrac{3.00 \times 10^8 \text{ m/s}}{107 \times 10^6 \text{ Hz}} = 2.80$ m.

The length of the antenna of a radio station is determined by the wavelength of the emitted radio waves. By knowing the wavelength we want, we can construct the antenna to the right length.

III. Mathematical Summary

Magnetic Flux	$\Phi = BA \cos \theta$ (20.1)	Defines the magnetic flux through an area, where θ is the angle between the normal to the plane of the loop and the magnetic field vector
Faraday's Law of Induction	$\varepsilon = -N \dfrac{\Delta \Phi}{\Delta t}$ (20.2)	Calculates the induced emf as the change in magnetic flux divided by the change in time.
Generator emf	$\varepsilon = \varepsilon_0 \sin \omega t$ (20.4) where $\varepsilon_0 = NBA\omega$	Expresses the generated emf as a function of time.
Back emf in a Motor	$\varepsilon_b = V - IR$ (20.6)	Computes the back emf in an electric motor.
Current, Voltages and Turn Ratios for a Transformer	$\dfrac{V_s}{V_p} = \dfrac{I_p}{I_s} = \dfrac{N_s}{N_p}$ (20.10)	Relates the current, voltage, and turns ratio in a transformer.

IV. Solutions of Selected Exercises and Paired Exercises

6. (a) When the bar magnet enters the coil, the needle deflects to one side, and when the magnet leaves the coil, the needle reverses direction.

   (b) $\boxed{\text{No}}$, by Lenz's law, it is repelled, moving toward the loop, and attracted as it leaves the loop.

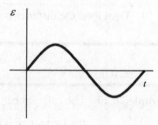

12. $\Phi = BA \cos \theta$, where θ is the angle between the field and the normal to the loop.

   (a) $\Phi = BA \cos 90° = \boxed{0}$.

   (b) $\Phi = (0.30 \text{ T})(0.015 \text{ m}^2) \cos (90° - 37°) = \boxed{2.7 \times 10^{-3} \text{ T·m}^2}$.

   (c) $\Phi = (0.30 \text{ T})(0.015 \text{ m}^2) \cos 0° = \boxed{4.5 \times 10^{-3} \text{ T·m}^2}$.

20. $\varepsilon = -N\dfrac{\Delta\Phi}{\Delta t} = -N\dfrac{A\Delta B}{\Delta t} = -(1)\times\dfrac{0.15\ \text{Wb} - 0.35\ \text{Wb}}{0.20\ \text{s}} = 1.0\ \text{V}.$

So $R = \dfrac{\varepsilon}{I} = \dfrac{1.0\ \text{V}}{10\ \text{A}} = \boxed{0.10\ \Omega}.$

26. $v = 320\ \text{km/h} = 88.9\ \text{m/s}.$ From Example 20.4 in the text on page 663, the magnitude of the induced emf is

$\varepsilon = BLv = (5.0\times10^{-5}\ \text{T})(30\ \text{m})(88.9\ \text{m/s}) = \boxed{0.13\ \text{V}}.$

30. (a) The answer is $\boxed{(1)\ \text{higher}}$. The change in magnetic flux is greater if it is flipped by $180°$ than by $90°$ because $\cos 0° - \cos 180° = 2$ and $\cos 0° - \cos 90° = 1$. A greater change in magnetic flux will result in a higher emf if the time interval is the same.

(b) $\varepsilon = -N\dfrac{\Delta\Phi}{\Delta t} = -(10)\times\dfrac{(1.8\ \text{T})(0.055\ \text{m}^2)(\cos 0° - \cos 180°)}{0.25\ \text{s}} = -\boxed{8.0\ \text{V}}.$

36. If the armature is jammed or turns very slowly, there is no back emf, and thus there is a large current.

40. (a) $\varepsilon = \varepsilon_0 \sin 2\pi f t = \varepsilon_0 \sin 2\pi(60\ \text{Hz})t = \boxed{\varepsilon_0 \sin 120\pi t}.$

(b) $\varepsilon_0 = NBA\omega = 2\pi NBAf = 2\pi(10)(0.350\ \text{T})(50\times10^{-4}\ \text{m}^2)(60\ \text{Hz}) = \boxed{6.6\ \text{V}}.$

44. $\varepsilon_0 = NBA\omega = 2\pi NBAf = 2\pi(20)(0.800\ \text{T})(\pi)(0.10\ \text{m})^2(60\ \text{Hz}) = \boxed{1.9\times10^2\ \text{V}}.$

This maximum value (positive or negative) is attained every half of a period.

$t = \dfrac{T}{2} = \dfrac{1}{2f} = \dfrac{1}{2(60\ \text{Hz})} = \boxed{\text{every } 1/120\ \text{s}}.$

48. (a) The back emf is $\varepsilon_b = V - IR$, so $I = \dfrac{V - \varepsilon_b}{R} = \dfrac{12\ \text{V} - 10\ \text{V}}{0.40\ \Omega} = \boxed{5.0\ \text{A}}.$

(b) When at half its final rotational speed, the back emf is also half (because induced emf is directly proportional to angular speed). $I = \dfrac{12\ \text{V} - 5.0\ \text{V}}{0.40\ \Omega} = 17.5\ \text{A} = \boxed{18\ \text{A}}.$

(c) When initially starting up, there is no back emf.

So $I = \dfrac{12\ \text{V}}{0.40\ \Omega} = \boxed{30\ \text{A}}.$

54. (a) This is a $\boxed{(1)\ \text{step–up}}$ transformer because there are more turns in the secondary, $N_s > N_p$.

(b) $\dfrac{I_p}{I_s} = \dfrac{V_s}{V_p} = \dfrac{N_s}{N_p} = \dfrac{450}{75} = \boxed{6{:}1}$.

(c) $\dfrac{V_p}{V_s} = \dfrac{N_p}{N_s} = \dfrac{75}{450} = \boxed{1{:}6}$.

60. (a) Because $\dfrac{I_s}{I_p} = \dfrac{V_p}{V_s}$, $e = \dfrac{I_s V_s}{I_p V_p} = \dfrac{V_p}{V_s}\dfrac{V_s}{V_p} = 1 = 100\%$, which implies that it is under ideal conditions and

there is no loss to the wires in the windings, etc.

(b) $e = \dfrac{I_s V_s}{I_p V_p} = \dfrac{(5.0\ \text{A})(240\ \text{V})}{(12\ \text{A})(120\ \text{V})} = 0.83 = \boxed{83\%}$. So $\boxed{\text{No}}$, it is not ideal.

62. (a) From the transformer equation, $\dfrac{V_s}{V_p} = = \dfrac{I_p}{I_s} = \dfrac{N_s}{N_p}$,

we have $N_s = \dfrac{V_s}{V_p} N_p = \dfrac{20\ \text{V}}{120\ \text{V}} \times (300) = \boxed{50}$ turns.

(b) $I_p = \dfrac{N_s}{N_p} I_s = \dfrac{50}{300} \times (0.50\ \text{A}) = \boxed{8.3 \times 10^{-2}\ \text{A}}$.

68. (a) $P_o = I^2 R = (50\ \text{A})^2 (1.2\ \Omega/\text{km})(25\ \text{km}) = \boxed{75\ \text{kW}}$.

(b) $P = \dfrac{P_o}{15} = \dfrac{75\,000\ \text{W}}{15} = 5000\ \text{W}$.

So $I = \sqrt{\dfrac{P}{R}} = \sqrt{\dfrac{5000\ \text{W}}{(1.2\ \Omega/\text{km})(25\ \text{km})}} = 12.9\ \text{A}$.

Power ($P = IV$) is conserved so $V_s = \dfrac{I_p}{I_s} V_p = \dfrac{50\ \text{A}}{12.9\ \text{A}} \times (20\,000\ \text{V}) = \boxed{77\ \text{kV}}$.

75. UV radiation causes sunburn and it can still go through the clouds. You feel cool because infrared (heat) radiation is absorbed by clouds (water molecules).

78. (a) From the wave equation, $c = \lambda f$, we have $f = \dfrac{c}{\lambda} = \dfrac{3.00 \times 10^8\ \text{m/s}}{0.030\ \text{m}} = \boxed{1.0 \times 10^{10}\ \text{Hz}}$.

(b) $f = \dfrac{3.00 \times 10^8\ \text{m/s}}{650 \times 10^{-9}\ \text{m}} = \boxed{4.6 \times 10^{14}\ \text{Hz}}$.

(c) $f = \dfrac{3.00 \times 10^8\ \text{m/s}}{1.2 \times 10^{-15}\ \text{m}} = \boxed{2.5 \times 10^{23}\ \text{Hz}}$.

(d) $\boxed{\text{(a): microwave, (b) visible light, (c) gamma ray}}$.

84.　(a) According to $c = \lambda f$, wavelength and frequency are inversely proportional to each other. So the longer

the distance between the cold spots, the longer is the wavelength, $\boxed{\text{(2) the lower is the frequency}}$.

(b) $\boxed{\text{Your microwave}}$ operates at a higher frequency.

Your microwave:　　　　　　　　　$\lambda = 2(5.0 \text{ cm}) = 10 \text{ cm} = 0.10 \text{ m}.$

$$f = \frac{c}{\lambda} = \frac{3.00 \times 10^8 \text{ m/s}}{0.10 \text{ m}} = 3.0 \times 10^9 \text{ Hz}.$$

Neighbor's microwave:　　　　　　$\lambda = 2(6.0 \text{ cm}) = 12 \text{ cm} = 0.12 \text{ m}.$

$$f = \frac{3.00 \times 10^8 \text{ m/s}}{0.12 \text{ m}} = 2.5 \times 10^9 \text{ Hz}.$$

The frequency difference is $3.0 \times 10^9 \text{ Hz} - 2.5 \times 10^9 \text{ Hz} = 0.5 \times 10^9 \text{ Hz} = \boxed{5.0 \times 10^8 \text{ Hz}}$.

86.　(a) From the power in the primary, $P_p = I_p V_p$, we have $I_p = \dfrac{P_p}{V_p} = \dfrac{1.00 \times 10^9 \text{ W}}{500 \text{ V}} = \boxed{2.00 \times 10^6 \text{ A}}$.

(b) The voltage ratio is $\dfrac{750 \times 10^3 \text{ V}}{500 \text{ V}} = 1500$. In five (5) steps, each transformer will step-up by a factor of

$\sqrt[5]{1500} = 4.32$ so $4.32^5 = 1500$.

Therefore the turn ratio of each transformer is $\boxed{N_p/N_s = 1 : 4.32}$.

(c) Since $\dfrac{I_s}{I_p} = \dfrac{V_p}{V_s}$, $I_s = \dfrac{V_p}{V_s} \times I_p = \dfrac{500 \text{ V}}{750 \times 10^3 \text{ V}} \times (2.00 \times 10^6 \text{ A}) = \boxed{1.33 \times 10^3 \text{ A}}$.

89.　(a) The induced current in the coil is $\boxed{\text{(1) clockwise}}$.

(b) $\varepsilon = IR = (10.0 \times 10^{-3} \text{ A})(0.500 \text{ }\Omega) = \boxed{5.00 \times 10^{-3} \text{ V}}$.

(c) $\varepsilon = -N \dfrac{\Delta \Phi}{\Delta t} = -N \dfrac{0 - BA}{\Delta t} = \dfrac{NBA}{\Delta t}$.

So $\Delta t = \dfrac{NBA}{\varepsilon} = \dfrac{(100)(3.50 \times 10^{-3} \text{ T})(\pi)(0.0200 \text{ m})^2}{5.00 \times 10^{-3} \text{ V}} = \boxed{0.0879 \text{ s}}$.

V.　Practice Quiz

1.　According to Lenz's law, the direction of an induced current in a conductor will be that which tends to

(a) enhance the effect which produces it.　　(b) oppose the effect that produces it.

(c) produce the greatest voltage.　　　　　　(d) produce a greater heating effect.

(e) none of the preceding

2.   A dc motor with an internal resistance of 10 Ω operating on a 12-V source produces a back emf of 9.0 V. What is the operating current through the motor?

     (a) zero   (b) 0.30 A   (c) 0.90 A   (d) 1.2 A   (e) 2.1 A

3.   A uniform magnetic field of 0.50 T passes perpendicularly through the plane of a wire loop 0.30 m^2 in area. What is the magnetic flux through the loop?

     (a) 1.7 Wb   (b) 0.80 Wb   (c) 0.60 Wb   (d) 0.15 Wb   (e) zero

4.   A 0.50-m-long wire is moved perpendicularly to a magnetic field of 0.30 T at a speed of 12 m/s. What motional emf is induced across the ends of the wire?

     (a) zero   (b) 0.15 V   (c) 1.8 V   (d) 3.6 V   (e) 6.0 V

5.   A bar magnet falls through a loop of wire with the south pole entering first. When the south pole enters the wire the induced current will be (as viewed from above)

     (a) clockwise.   (b) counterclockwise.   (c) zero.

     (d) to the right of the loop.   (e) to the left of the loop.

6.   A coil consists of five loops of wire, each with an area of 0.20 m^2. The coil is oriented with its plane perpendicular to a magnetic field that increases uniformly from 1.0×10^{-2} T to 2.5×10^{-2} T in a time of 5.0×10^{-3} s. What is the induced emf in the coil?

     (a) zero   (b) 0.60 V   (c) 2.0 V   (d) 3.0 V   (e) 5.0 V

7.   You are designing a generator with a maximum emf of 24 V. If the generator coil has 100 turns and a cross-sectional area of 0.030 m^2, what will be the required magnetic field if the frequency is 60 Hz?

     (a) zero   (b) 0.022 T   (c) 0.042 T   (d) 0.067 T   (e) 0.13 T

8.   A transformer consists of a 500-turn primary coil and a 2000-turn secondary coil. If the voltage in the secondary is 4.8 V, what is the primary voltage?

     (a) 0.30 V   (b) 0.60 V   (c) 1.2 V   (d) 2.4 V   (e) 19 V

9.   An electromagnetic wave is made up of which of the following time-varying quantities?

     (a) electrons only   (b) electric fields only   (c) magnetic fields only

     (d) both electric and magnetic fields   (e) neither electric fields nor magnetic fields

10.   What is the frequency of a radio-wave signal transmitted at a wavelength of 40 m?

     (a) 7.5 Hz   (b) 750 Hz   (c) 7.5 kHz   (d) 75 KHz   (e) 7.5 MHz

11. The transformer on a utility pole steps the voltage down from 20 000 V to 220 V for use in a subdivision. If the subdivision uses 200 kW of power, what is the current in the primary coil?

(a) 1.0 A   (b) 10 A   (c) 91 A   (d) 910 A   (e) none of the preceding

12. A certain gamma ray used in medical treatment has a frequency of about 1×10^{23} Hz. What is its wavelength?

(a) 1×10^{23} m   (b) 1×10^{15} m   (c) 3×10^8 m   (d) 3×10^{-15} m   (e) none of the preceding

Answers to Practice Quiz:

1.b 2.b 3.d 4.c 5.a 6.d 7.b 8.c 9.d 10.e 11.b 12.d

CHAPTER 21

AC Circuits

I.   Chapter Objectives

Upon completion of this chapter, you should be able to:

1.   specify how voltage, current, and power vary with time in an ac circuit, understand the concepts of rms and peak values, and learn how resistors respond under ac conditions.

2.   explain the behavior of capacitors in ac circuits, and calculate the effect of a capacitor on ac current (capacitive reactance).

3.   explain what an inductor is, explain the behavior of inductors in ac circuits, and calculate the effect of inductors on ac circuits (inductive reactance).

4.   calculate currents and voltages when a combination of reactive and resistive circuit elements are present in ac circuits, use phase diagrams to calculate overall impedance and rms currents, and understand and use the concept of the power factor in ac circuits.

5.   understand the concept of resonance in ac circuits, and calculate the resonance frequency of an RLC circuit.

II.   Chapter Summary and Discussion

1.   Resistance in an AC Circuit (Section 21.1)

In an ac circuit, voltage and current vary sinusoidally (or voltages and currents are sine or cosine functions) with time. The instantaneous voltage is given by $V = V_0 \sin \omega t = V_0 \sin 2\pi f t$, where V_0 is the **peak voltage**, which is the maximum voltage value attained during a cycle of oscillation, and $\omega = 2\pi f$ is the angular frequency.

In an ac circuit with one resistor, the current through that resistor is equal to $I = \dfrac{V}{R} = \dfrac{V_0 \sin 2\pi f t}{R}$

$= \dfrac{V_0}{R} \sin 2\pi f t = I_0 \sin 2\pi f t$, where I_0 is the **peak current** (maximum current value). The voltage across and current through the resistor are in step, or *in phase*, with each other, both reaching zero, minima, and maxima at the same time.

The voltage and current in ac circuits are often given in terms of the **rms voltage** and **rms current**, unless otherwise specified. These are statistical averages of the **effective** *ac voltage and current*. The rms values are less than the peak values and are related to the peak values by $V_{rms} = \dfrac{V_o}{\sqrt{2}} \approx 0.707 V_o$, and $I_{rms} = \dfrac{I_o}{\sqrt{2}} \approx 0.707 I_o$. The rms voltage and current are related by $V_{rms} = I_{rms}R$, and the average power dissipated by a resistor is equal to $\overline{P} = I_{rms}V_{rms} = I_{rms}^2 R$. Appliances using ac voltage are rated in terms of their average power.

Example 21.1   An electric hot plate rated at 200 W is connected to a 120-V outlet.

(a) What are the rms and peak currents through the hot plate?

(b) What is the resistance of the hot plate?

Solution: Given: $\overline{P} = 200$ W, $V_{rms} = 120$ V.

Find: (a) I_{rms} and I_o (b) R.

(a) From $\overline{P} = V_{rms}I_{rms}$, we have $I_{rms} = \dfrac{\overline{P}}{V_{rms}} = \dfrac{200 \text{ W}}{120 \text{ V}} = 1.67$ A.

The peak current is then $I_o = \sqrt{2}\, I_{rms} = \sqrt{2}\,(1.67 \text{ A}) = 2.36$ A.

(b) $R = \dfrac{V_{rms}}{I_{rms}} = \dfrac{120 \text{ V}}{1.67 \text{ V}} = 71.9\ \Omega$.

Example 21.2   An ac appliance of 10-Ω resistance is connected to a 120-V ac power supply.

(a) What is the peak current through the appliance?

(b) What is the average power dissipated in the appliance?

Solution: Given: $R = 10\ \Omega$, $V_{rms} = 120$ V.

Find: (a) I_o (b) $\overline{P}$.

(a) $I_{rms} = \dfrac{V_{rms}}{R} = \dfrac{120 \text{ V}}{10\ \Omega} = 12$ A.

Therefore, the peak current is $I_o = \sqrt{2}\, I_{rms} = \sqrt{2}\,(12 \text{ A}) = 17$ A.

(b) $\overline{P} = I_{rms}V_{rms} = (12 \text{ A})(120 \text{ V}) = 1.44 \times 10^3$ W $= 1.4$ kW.

---

2. Capacitive Reactance (Section 21.2)

A capacitor in an ac circuit limits, or impedes, the current but does not completely prevent the flow of charge, as it does in dc circuits. This impeding effect is expressed in terms of **capacitive reactance** X_C given by $X_C = \dfrac{1}{\omega C} = \dfrac{1}{2\pi fC}$. The SI unit of capacitive reactance is the ohm (Ω). Capacitive reactance is frequency dependent (inversely proportional to frequency) and is analogous to resistance in the sense that the rms voltage across, and the rms current through, the capacitor are related by a more general form of Ohm's law, $V_{rms} = I_{rms}X_C$.

In a purely capacitive ac circuit (the capacitor is the only circuit element), the *current leads the voltage* by 90° or a quarter (1/4) of a cycle. This means the voltage and current are *not* in step, or *not* in phase. When the current reaches maximum, the voltage is zero, and when the current reaches zero, the voltage is maximum. Alternatively, we can say that the voltage lags the current by 90°. A capacitor simply charges and discharges in an ac circuit and does not dissipate power or consume energy.

Example 21.3   A capacitor of 10.0 μF is connected to an ac source of 120 V and 60 Hz.

       (a) What is the capacitive reactance?

       (b) What is the rms current?

       (c) What power does the capacitor dissipate?

Solution: Given: $C = 10.0\ \mu\text{F} = 10.0 \times 10^{-6}$ F, $V = 120$ V, $f = 60$ Hz (assume exact).

            Find: (a) X_C (b) I (c) P.

(a) $X_C = \dfrac{1}{2\pi fC} = \dfrac{1}{2\pi(60\ \text{Hz})(10.0 \times 10^{-6}\ \text{F})} = 265\ \Omega$.

(b) From Ohm's law, $I_{rms} = \dfrac{V_{rms}}{X_C} = \dfrac{120\ \text{V}}{265\ \Omega} = 0.453$ A.

(c) An ideal capacitor ($R = 0$) does not dissipate power, so $P = 0$.

---

3. Inductive Reactance (Section 21.3)

An inductor in an ac circuit also limits, or impedes, the current but does not completely prevent the flow of charge. This impeding effect is expressed in terms of **inductive reactance** X_L, given by $X_L = \omega L = 2\pi fL$, where L is the **inductance** of the inductor and has the SI unit **henry** (H). The SI unit of inductive reactance X_L is also the ohm (Ω). Inductive reactance is frequency dependent (directly proportional to frequency) and is analogous to resistance in the sense that the rms voltage across, and the rms current through, the inductor are related by a more general form of Ohm's law, $V_{rms} = I_{rms}X_L$.

In a purely inductive ac circuit (the inductor is the only circuit element), the *voltage leads the current* by 90° or a quarter (1/4) of a cycle. This means the voltage and current are *not* in step, or *not* in phase. When the voltage reaches maximum, the current is zero, and when the voltage reaches zero, the current is maximum. Alternatively, the current lags the current by 90°. A inductor simply stores and releases magnetic energy in an ac circuit and does not dissipate power or consume energy.

Note: The phase relationship of current and voltage for purely inductive and purely capacitive circuits are the opposite of each other. A phrase that may help you remember the relationship is *EL̲I* the *IC̲E* man. With *E* representing voltage and *I* representing current, *EL̲I* indicates that with an inductance (*L*) the voltage leads the current (*I*). Similarly, *IC̲E* tells you that with a capacitance (*C*) the current leads the voltage.

Example 21.4   An inductor of 25.0 mH is connected to an ac source of 120 V and 60 Hz. Find the inductive reactance and the rms current in the circuit.

Solution: Given: $L = 25.0$ mH $= 25.0 \times 10^{-3}$ H, $V = 120$ V, $f = 60$ Hz.

Find: X_L and I.

The inductive reactance is $X_L = 2\pi fL = 2\pi (60 \text{ Hz})(25.0 \times 10^{-3} \text{ H}) = 9.42 \ \Omega$.

From Ohm's law, the rms current is equal to $I_{rms} = \dfrac{V_{rms}}{X_L} = \dfrac{120 \text{ V}}{9.42 \ \Omega} = 12.7$ A.

---

4.   Impedance: RLC Circuits (Section 21.4)

Impedance Z is a generalization of opposition to current that includes not only resistance but also capacitive and inductive reactances. Because of the phase difference between the voltage and currents on different circuit elements, a **phase diagram** is convenient to use. In a phase diagram, the resistance and reactance of the circuit are given vectorlike properties and their magnitudes are represented as arrows called **phasors**. On a set of *x*-*y* axes, the resistance is used as a reference and is plotted on the positive *x*-axis, since the voltage-current phase difference for a resistor is zero ($\phi = 0$). The reactances are plotted based on their phase differences. For example, the voltage lags the current by 90° in a capacitor, so the capacitive reactance is plotted on the negative *y*-axis; the voltage leads the current by 90° for a inductor, so the inductive reactance is plotted on the positive *y*-axis.

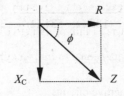 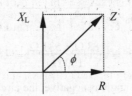

From the phase diagram, the impedance in an RLC series circuit is given by

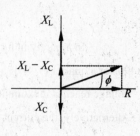

$$Z = \sqrt{R^2 + (X_L - X_C)^2} = \sqrt{R^2 + \left(2\pi fL - \frac{1}{2\pi fC}\right)^2}.$$ ϕ is called the **phase angle** of the

circuit and is equal to $\tan\phi = \dfrac{X_L - X_C}{R}$. If $X_L > X_C$, the angle ϕ is positive and the circuit

is said to be an **inductive circuit**; if $X_C > X_L$, the angle ϕ is negative and the circuit is said to be a **capacitive circuit**.

If the circuit is an RC circuit (no inductor), then $X_L = 0$, $Z = \sqrt{R^2 + X_C^2}$ and $\tan\phi = -\dfrac{X_C}{R}$; if the circuit

is an RL circuit (no capacitor), then $X_C = 0$, $Z = \sqrt{R^2 + X_L^2}$, and $\tan\phi = \dfrac{X_L}{R}$. The general form of Ohm's law

relating the rms voltage to the rms current, $V_{rms} = I_{rms}Z$, is valid for all such circuits.

In an RLC series circuit, *there are no power losses associated with a pure capacitor and inductor. Only the resistor dissipates power.* Therefore, the average power in an ac circuit can be expressed as $\overline{P} = I_{rms}V_{rms}\cos\phi$

$= I_{rms}^2 Z \cos\phi$, where the term $\cos\phi$ is called the **power factor** and is defined as $\cos\phi = \dfrac{R}{Z}$ from the phase diagram.

The power factor is a measure of how close the circuit is to expending the maximum power ($\cos 0° = 1$).

Example 21.5   An RLC series circuit consists of a resistor of 100 Ω, a capacitor of 5.00 μF, and an inductor of 0.500 H. The circuit is connected to a power supply of 120 V operating at 60 Hz.

(a) What is the impedance?

(b) What is the phase angle? Is the circuit inductive or capacitive?

(c) What is the rms current?

(d) What is the rms voltage across each element?

(e) What is the average power dissipated in the circuit?

Solution: Given: $R = 100\ \Omega$, $C = 5.00\ \mu F = 5.00 \times 10^{-6}$ F, $L = 0.500$ H, $V_{rms} = 120$ V, $f = 60$ Hz.

Find: (a) Z (b) ϕ (c) I_{rms} (d) V_{rms} across each element   (e) $\overline{P}$.

(a) $X_L = 2\pi fL = 2\pi(60$ Hz$)(0.500$ H$) = 188\ \Omega$, and

$$X_C = \frac{1}{2\pi fC} = \frac{1}{2\pi(60 \text{ Hz})(5.00 \times 10^{-6} \text{ F})} = 531\ \Omega,$$

so the impedance is $Z = \sqrt{R^2 + (X_L - X_C)^2} = \sqrt{(100\ \Omega)^2 + (188\ \Omega - 531\ \Omega)^2} = 357\ \Omega.$

(b) $\tan\phi = \dfrac{X_L - X_C}{R} = \dfrac{188\ \Omega - 531\ \Omega}{100\ \Omega} = -3.43,$ so $\phi = -73.7°.$

Because the phase angle is negative, the circuit is a capacitive circuit.

(c) $I_{rms} = \dfrac{V_{rms}}{Z} = \dfrac{120 \text{ V}}{357 \text{ } \Omega} = 0.336 \text{ A}$.

(d) $(V_{rms})_R = I_{rms}R = (0.336 \text{ A})(100 \text{ } \Omega) = 33.6 \text{ V}$, $(V_{rms})_C = I_{rms}X_C = (0.336 \text{ A})(531 \text{ } \Omega) = 178 \text{ V}$, and

$(V_{rms})_L = I_{rms}X_L = (0.336 \text{ A})(188 \text{ } \Omega) = 63.2 \text{ V}$. **Note:** $(V_{rms})_R + (V_{rms})_C + (V_{rms})_L = 33.6 \text{ V} + 178 \text{ V} + 63.2 \text{ V}$

$= 275 \text{ V} > 120 \text{ V}$. Is this a violation of the conservation of energy? How can the sum of the individual voltages be greater than the total voltage? (*Hint*: The voltages calculated here are rms, or effective voltages, and at any instant, the voltages are *not* in phase!)

(e) Only the resistor dissipates energy. $\overline{P} = I_{rms}^2 R = (0.336 \text{ A})^2 (100 \text{ } \Omega) = 11.3 \text{ W}$.

Or $\overline{P} = I_{rms}V_{rms} \cos \phi = (0.336 \text{ A})(120 \text{ V}) \cos 73.7° = 11.3 \text{ W}$.

5. Circuit Resonance (Section 21.5)

When $X_L = X_C$ (the capacitive reactance and the inductive reactance cancel each other completely), $\phi = 0$ and $Z = R$ (completely resistive) is at minimum, so the circuit dissipates maximum power. This condition is called *resonance*, and the frequency at which resonance occurs is called the **resonance frequency** (f_o). At resonance,

$X_L = X_C$, or $2\pi f_o L = \dfrac{1}{2\pi f_o C}$, so $f_o = \dfrac{1}{2\pi \sqrt{LC}}$. Resonant circuits have a wide variety of applications such as radio and television tuning.

Integrated Example 21.6

An RLC series circuit consists of a resistor of 100 Ω, a capacitor of 20.0 μF, and an inductor of 0.550 H. The circuit is connected to a power supply of 120 V with an adjustable (tunable) frequency and tuned to resonance. (a) The rms voltage across the capacitor is (1) higher than, (2) the same as, or (3) lower than the rms voltage across the inductor. Explain. (b) Calculate the resonant frequency, the impedance, the rms current, the average power, and the rms voltages across all three circuit elements.

(a) Conceptual Reasoning:

The concept of resonance corresponds to a situation under which the impedance is at minimum. Since impedance is $Z = \sqrt{R^2 + (X_L - X_C)^2}$. It is at minimum when $X_L = X_C$. Since the three circuit elements are in series, they have the same rms current. Therefore, when $X_L = X_C$, the rms voltage across the capacitor is (2) the same as the rms voltage across the inductor.

(b) Quantitative Reasoning and Solution:

Given: $R = 100 \text{ } \Omega$, $C = 20.0 \mu\text{F} = 20.0 \times 10^{-6} \text{ F}$, $L = 0.550 \text{ H}$, $V_{rms} = 120 \text{ V}$.

Find: f_o, Z, I_{rms}, $\overline{P}$, and V_{rms} for each element.

$$f_o = \frac{1}{2\pi\sqrt{LC}} = \frac{1}{2\pi\sqrt{(20.0 \times 10^{-6}\text{ F})(0.550\text{ H})}} = 48.0\text{ Hz}.$$

At resonance, $X_L = X_C$, so $Z = \sqrt{R^2 + (X_L - X_C)^2} = R = 100\ \Omega$ (it is at minimum).

$$I_{\text{rms}} = \frac{V_{\text{rms}}}{Z} = \frac{120\text{ V}}{100\ \Omega} = 1.20\text{ A}.$$

$$\overline{P} = I_{\text{rms}}^2 R = (1.20\text{ A})^2(100\ \Omega) = 144\text{ W}.$$

$$(V_{\text{rms}})_R = I_{\text{rms}}R = (1.20\text{ A})(100\ \Omega) = 120\text{ V}.$$

$$X_L = 2\pi fL = 2\pi(48.0\text{ Hz})(0.550\text{ H}) = 166\ \Omega, \text{ so } (V_{\text{rms}})_L = I_{\text{rms}}X_L = (1.20\text{ A})(166\ \Omega) = 199\text{ V}.$$

$$X_C = \frac{1}{2\pi fC} = \frac{1}{2\pi(48.0\text{ Hz})(20.0 \times 10^{-6}\text{ F})} = 166\ \Omega, \text{ so } (V_{\text{rms}})_C = I_{\text{rms}}X_C = (1.20\text{ A})(166\ \Omega) = 199\text{ V}.$$

As expected in (a), the rms voltage across the inductor is the same as the rms voltage across the capacitor, $(V_{\text{rms}})_L = (V_{\text{rms}})_C$. However, these two voltages are out of phase (180°), so they cancel out.

III.   Mathematical Summary

Instantaneous ac Voltage	$V = V_0 \sin \omega t = V_0 \sin 2\pi ft$ (21.1)	Gives the instantaneous voltage in an ac circuit as a function of time.
Rms (or Effective) Current	$I_{\text{rms}} = \dfrac{I_0}{\sqrt{2}} \approx 0.707 I_0$ (21.6)	Relates the rms current and the peak current.
Rms (or Effective) Voltage	$V_{\text{rms}} = \dfrac{V_0}{\sqrt{2}} \approx 0.707 V_0$ (21.8)	Relates the rms voltage and the peak voltage.
Average Power in an ac Circuit	$\overline{P} = I_{\text{rms}}^2 R$ (21.10)	Computes the effective (average) power in an ac circuit.
Capacitive Reactance	$X_C = \dfrac{1}{\omega C} = \dfrac{1}{2\pi fC}$ (21.11)	Defines the capacitive reactance of a capacitor in an ac circuit.
Inductive Reactance	$X_L = \omega L = 2\pi fL$ (21.14)	Defines the inductive reactance of an inductor in an ac circuit.
Ohm's Law (Resistor)	$V_{\text{rms}} = I_{\text{rms}}R$ (21.9)	Gives the relationship between the rms voltages and rms current for a resistor.
Ohm's Law (Capacitor)	$V_{\text{rms}} = I_{\text{rms}}X_C$ (21.12)	Gives the relationship between the rms voltage and rms current for a capacitor.
Ohm's Law (Inductor)	$V_{\text{rms}} = I_{\text{rms}}X_L$ (21.15)	Gives the relationship between the rms voltage and rms current for an inductor.

Ohm's Law Generalized to ac Circuits	$V_{rms} = I_{rms} Z$ (21.17)	Relates rms voltage and rms current in any ac circuit (R, RC, RL, or RLC series).
Impedance for a Series RLC Circuit	$Z = \sqrt{R^2 + (X_L - X_C)^2}$ (21.19)	Defines the impedance of an RLC series circuit.
Phase Angle between Voltage and Current in a Series RLC Circuit	$\tan \phi = \dfrac{X_L - X_C}{R}$ (21.20)	Defines the phase angle between the voltage and current in an ac circuit.
Power Factor for an RLC Circuit	$\cos \phi = \dfrac{R}{Z}$ (21.22)	Defines the power factor in an ac circuit.
Average Power in Terms of Power Factor	$\bar{P} = I_{rms} V_{rms} \cos \phi$ (21.23) or $\bar{P} = I_{rms}^2 Z \cos \phi$ (21.24)	Expresses the average power in an ac circuit in terms of the power factor.
Resonance Frequency of a Series RLC Circuit	$f_o = \dfrac{1}{2\pi \sqrt{LC}}$ (21.25)	Computes the resonance frequency in an RC or RLC series circuit.

IV. Solutions of Selected Exercises and Paired Exercises

5. That means the voltage and current both reach maximum or minimum at the same time.

12. (a) $V_{rms} = I_{rms} R = (0.75 \text{ A})(5.0 \ \Omega) = \boxed{3.8 \text{ V}}$;

$V_o = \sqrt{2} \ V_{rms} = \sqrt{2} \ (3.75 \text{ V}) = \boxed{5.3 \text{ V}}$.

(b) $\bar{P} = I_{rms} V_{rms} = (0.75 \text{ A})(3.75 \text{ V}) = \boxed{2.8 \text{ W}}$.

16. (a) There are $\boxed{(2) \text{ two}}$ possible answers, because the voltage could be increasing or decreasing at the $t = 0$ moment. In other words, it could be either positive or negative since the direction is not specified.

(b) $T = \dfrac{1}{f} = \dfrac{1}{60 \text{ Hz}} = 1/60$ s so 1/240 s is ¼ the period. That is, the voltage at 1/240 s is either at the maximum or the minimum (negative side). Therefore, $V_{1/240 \text{ s}} = \boxed{\pm 85 \text{ V}}$.

18. From the average power , $\bar{P} = \dfrac{V_{rms}^2}{R}$, we have $R = \dfrac{V_{rms}^2}{\bar{P}} = \dfrac{(120 \text{ V})^2}{100 \text{ W}} = \boxed{144 \ \Omega}$.

$I_{rms} = \dfrac{V_{rms}}{R} = \dfrac{120 \text{ V}}{144 \ \Omega} = \boxed{0.833 \text{ A}}$.

27. For a capacitor, the *lower the frequency*, the longer the charging time in each cycle. If the frequency is very low (dc), then the charging time is very long, so it acts as an ac open circuit.

For an inductor, the *lower the frequency*, the more slowly the current changes in the inductor The more slowly the current changes, the less back emf is induced in the inductor, resulting in less impedance to current.

30. From the capacitive reactance, $X_C = \dfrac{1}{2\pi f C}$, we have

$$f = \frac{1}{2\pi C X_C} = \frac{1}{2\pi (25 \times 10^{-6}\,\text{F})(25\,\Omega)} = \boxed{2.5 \times 10^2\,\text{Hz}}.$$

36. From the inductive reactance, $X_L = 2\pi f L$, we have

$$L = \frac{X_L}{2\pi f} = \frac{90\,\Omega}{2\pi(60\,\text{Hz})} = \boxed{0.24\,\text{H}}.$$

42. (a) The inductive reactance will be 1/2 times the original reactance because X_L is proportional to frequency, $X_L = 2\pi f L$. Then, according to Ohm's law $I_{rms} = V_{rms}/X_L$, the rms current is inversely proportional to the frequency. Therefore the rms current will be $\boxed{(1)\ 2}$ times the original rms current.

(b) $\dfrac{(I_{rms})_2}{(I_{rms})_1} = \dfrac{f_1}{f_2} = \dfrac{40\,\text{Hz}}{120\,\text{Hz}} = 1/3.$

So $(I_{rms})_2 = (1/3)(I_{rms})_1 = (1/3)(9.0\,\text{A}) = \boxed{3.0\,\text{A}}.$

50. (a) $X_C = \dfrac{1}{2\pi f C} = \dfrac{1}{2\pi(60\,\text{Hz})(25 \times 10^{-6}\,\text{F})} = \boxed{1.1 \times 10^2\,\Omega}.$

$Z = \sqrt{R^2 + (X_L - X_C)^2} = \sqrt{(200\,\Omega)^2 + (0 - 106\,\Omega)^2} = 226\,\Omega = \boxed{2.3 \times 10^2\,\Omega}.$

(b) $I_{rms} = \dfrac{V_{rms}}{Z} = \dfrac{120\,\text{V}}{226\,\Omega} = \boxed{0.53\,\text{A}}.$

56. (a) You can change $\boxed{(1)\ \text{resistance}}$ without upsetting the condition of resonance. At resonance, $X_L = X_C$. So as long as that condition is maintained, the condition satisfies.

(b) $X_C = \dfrac{1}{2\pi f C} = \dfrac{1}{2\pi(60\,\text{Hz})(3.5 \times 10^{-6}\,\text{F})} = 758\,\Omega.$

At resonance, $X_L = X_C = 2\pi f L$. So $L = \dfrac{758\,\Omega}{2\pi(60\,\text{Hz})} = \boxed{2.0\,\text{H for any resistance}}.$

58. At resonance, $f_0 = \dfrac{1}{2\pi\sqrt{LC}} = \dfrac{1}{2\pi\sqrt{(0.100 \text{ H})(5.00 \times 10^{-6} \text{ F})}} = \boxed{225 \text{ Hz}}$.

64. (a) $Z = \sqrt{R^2 + (X_L - X_C)^2} = \sqrt{(10 \ \Omega)^2 + (120 \ \Omega - 120 \ \Omega)^2} = 10 \ \Omega$.

$I_{rms} = \dfrac{V_{rms}}{Z} = \dfrac{220 \text{ V}}{10 \ \Omega} = 22$ A. So $(V_{rms})_R = I_{rms}R = (22 \text{ A})(10 \ \Omega) = \boxed{220 \text{ V}}$.

(b) $(V_{rms})_L = I_{rms}X_L = (22 \text{ A})(120 \ \Omega) = \boxed{2.64 \times 10^3 \text{ V}}$.

(c) $(V_{rms})_C = I_{rms}X_C = (22 \text{ A})(120 \ \Omega) = \boxed{2.64 \times 10^3 \text{ V}}$.

67. (a) The impedance is $\boxed{(2) \text{ equal to } 25 \ \Omega}$. At resonance, $X_L = X_C$, so $Z = R$.

(b) $X_L = 2\pi fL = 2\pi (60 \text{ Hz})(0.450 \text{ H}) = 170 \ \Omega$, $X_C = \dfrac{1}{2\pi fC} = \dfrac{1}{2\pi (60 \text{ Hz})(5.00 \times 10^{-6} \text{ F})} = 531 \ \Omega$.

$Z = \sqrt{R^2 + (X_L - X_C)^2} = \sqrt{(25.0 \ \Omega)^2 + (170 \ \Omega - 531 \ \Omega)^2} = \boxed{362 \ \Omega}$.

73. (a) The phase angle is $\boxed{(2) \text{ zero, as } X_L = X_C}$, according to $\phi = \tan^{-1}\dfrac{X_L - X_C}{R}$.

(b) From the resonant frequency, $f_0 = \dfrac{1}{2\pi\sqrt{LC}}$, we have

$C = \dfrac{1}{4\pi^2 f_0^2 L} = \dfrac{1}{4\pi^2 (60 \text{ Hz})(0.750 \text{ H})} = \boxed{9.4 \ \mu\text{F}}$.

V. Practice Quiz

1. The inductive reactance in an ac circuit will change by what factor when the frequency is tripled?

(a) 1/3   (b) 1/9   (c) 1   (d) 9   (e) 3

2. What is the peak voltage in an ac circuit with an rms voltage of 120 V?

(a) zero   (b) 84.9 V   (c) 120 V   (d) 170 V   (e) 240 V

3. What is the impedance of an ac circuit with a resistance of 12.0 Ω, an inductive reactance of 15.0 Ω, and a capacitive reactance of 10.0 Ω?

(a) 37.0 Ω (b) 27.7 Ω (c) 11.6 Ω (d) 21.9 Ω (e) 13.0 Ω

4. A RLC series circuit has a resistance of 12.0 Ω, an inductive reactance of 15.0 Ω, and a capacitive reactance of 10.0 Ω. If an rms voltage of 120 V is applied, what is the power output?

(a) 435 W   (b) 598 W   (c) 1022 W   (d) 1108 W   (e) 1200 W

5. What is the phase difference between the voltages of the inductor and capacitor in an RLC series circuit.

   (a) 45°   (b) 90°   (c) 180°   (d) 360°   (e) zero

6. What is the resonance frequency for a circuit containing an inductor of 10 mH and a capacitor of 20 μF?

   (a) zero   (b) 0.36 kHz   (c) 1.3 kHz   (d) 2.2 kHz   (e) 0.80 MHz.

7. An ac series circuit has an impedance of 60 Ω and a resistance of 30 Ω. What is the power factor?

   (a) 0.50   (b) 0.71   (c) 1.0   (d) 1.4   (e) 2.0

8. Resonance occurs in an ac series circuit when

   (a) resistance equals capacitive reactance.   (b) resistance equals inductive reactance.

   (c) capacitive reactance equals zero.   (d) capacitive reactance equals inductive reactance.

   (e) inductive reactance equals zero.

9. A 10-μF capacitor is connected to a 120-V, 60-Hz source. What is the rms current in the capacitor?

   (a) 0.072 A   (b) 0.45 A   (c) 3.2×10^4 A   (d) 2.0×10^5 A   (e) 1.2×10^7 A

10. A 60-W lightbulb operates on 120-V household voltage. What is the peak current in the bulb?

    (a) 0.51 A   (b) 0.71 A   (c) 1.4 A   (d) 2.0 A   (e) 2.8 A

11. An RLC series circuit consists of a resistor of 100 Ω, a capacitor of 10.0 μF, and an inductor of 0.250 H. The circuit is connected to a power supply of 120 V and 60 Hz. What is the power dissipated in the circuit?

    (a) zero   (b) 37 W   (c) 63 W   (d) 73 W   (e) 0.14 kW

12. What are the phase angle and power factor for an RLC series circuit containing a resistor of 50 Ω, a capacitor of 10 μF, and an inductor of 0.45 H, when connected to a power supply with frequency of 60 Hz?

    (a) −62° and 0.47   (b) 62° and 0.47   (c) −38° and 0.79   (d) 38° and 0.79   (e) zero and 1.0

Answers to Practice Quiz:

1.e 2.d 3.e 4.c 5.c 6.b 7.a 8.d 9.b 10.b 11.d 12.a

CHAPTER 22

Reflection and Refraction of Light

I. Chapter Objectives

Upon completion of this chapter, you should be able to:

1. define and explain the concept of wave fronts and rays.
2. explain the law of reflection, and distinguish between regular (specular) and irregular (diffuse) reflections.
3. explain refraction in terms of Snell's law and the index of refraction, and give examples of refractive phenomena.
4. describe total internal reflection, and understand fiber-optic applications.
5. explain dispersion and some of its effects.

II. Chapter Summary and Discussion

1. Wave Fronts and Rays (Section 22.1)

A **wave front** is the line (in two dimensions) or surface (in three dimensions) defined by adjacent portions of a wave that are in phase. For example, a point light source emits spherical wave fronts because the points having the same phase angle are on the surface of a sphere. For a parallel beam of light, the wave front is a **plane wave front**. The distance between adjacent wave fronts is the wavelength of the wave.

A **ray** is a line drawn perpendicular to a series of wave fronts and pointing in the direction of propagation of the wave. For a spherical wave, the rays are radially outward, and for a plane wave, they are parallel to one another. The use of wave fronts and rays in describing optical phenomena such as reflection and refraction is called **geometrical optics**.

2. Reflection (Section 22.2)

The **law of reflection** states that the **angle of incidence** (the angle between the incident ray and the normal) equals the **angle of reflection** (the angle between the reflected ray and the normal), $\theta_i = \theta_r$, where θ_i is the angle of incidence, and θ_r is the angle of reflection. The incident ray, the normal, and the reflected ray are always in the same plane.

Note: All angles are measured *from the normal* (*line perpendicular to the reflecting surface*).

Regular (specular) reflection occurs from smooth surfaces, with the reflected rays parallel to one another. **Irregular (diffuse) reflection** occurs from rough surfaces, with the reflected rays being at different angles.

Example 22.1   Two mirrors make an angle of 90° with each other. A ray is incident on mirror M_1 at an angle of 30° to the normal. Find the direction of the ray after it is reflected from mirrors M_1 and M_2.

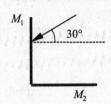

Solution:

All angles should be measured from the normal to the reflecting surface. The angle of incidence of the ray at M_1 is 30°. According to the law of reflection, the angle of reflection at M_1 is also 30°.

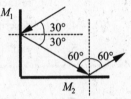

From geometry, the angle of incidence at M_2 is 90° − 30° = 60°. Therefore, the angle of reflection at M_2 is also 60°, and the ray emerges parallel to the original incident ray.

---

3.   Refraction (Section 22.3)

Refraction refers to the change in direction of a wave at a boundary where it passes from one medium into another as a result of different wave speeds in different media.

Snell's law relates the angle of incidence, θ_1, and angle of refraction θ_2, (the angle between the refracted ray and the normal) to the wave speeds in the respective media: $\dfrac{\sin \theta_1}{\sin \theta_2} = \dfrac{v_1}{v_2}$. The **index of refraction** of a medium is defined as the ratio of the speed of light in vacuum to its speed in that medium, $n = c/v$. Snell's law can be conveniently expressed in terms of the indices of refraction: $n_1 \sin \theta_1 = n_2 \sin \theta_2$. If the second medium is more optically dense $(n_2 > n_1)$, then $\theta_1 > \theta_2$, or the refracted ray is bent toward the normal; if the second medium is less dense $(n_2 < n_1)$, then $\theta_2 > \theta_1$, and the refracted ray is bent away from the normal.

When light travels from one medium to another the frequency remains constant, but the speed and wavelength change. In terms of wavelength, the index of refraction can be rewritten as $n = \lambda/\lambda_m$, where λ is the wavelength in vacuum, and λ_m is the wavelength in the medium.

Note: Because the geometrical representation of light is used here, a diagram is very helpful (if not necessary) in solving problems. Again, all angles are measured from the normal to the media boundary.

Example 22.2   A light ray travels through an air–fused quartz interface at an angle of 30° to the normal. Find the speed of light in the quartz and the angle of refraction.

Solution: Given: $n_2 = 1.46$ (fused quartz, from Table 22.1),

$n_1 \approx 1.00$ (air), $\theta_1 = 30°$.

Find: v_2 and θ_2.

From $n = \dfrac{c}{v}$, we have $v_2 = \dfrac{c}{n_2} = \dfrac{3.00 \times 10^8 \text{ m/s}}{1.46} = 2.05 \times 10^8 \text{ m/s}$.

From Snell's law, $n_1 \sin \theta_1 = n_2 \sin \theta_2$, we have $\sin \theta_2 = \dfrac{n_1 \sin \theta_1}{n_2} = \dfrac{(1.00) \sin 30°}{1.46} = 0.342$.

So $\theta_2 = \sin^{-1}(0.342) = 20°$

Integrated Example 22.3

A beam of light traveling in medium 1 is incident on medium 2 that is a slab of transparent material. The incident beam and the refracted beam make angles of 29° and 26° to the normal, respectively. (a) The index of medium 2 is (1) greater than, (2) equal to, or (3) less than the index of refraction of medium 1. Explain. (b) If medium 1 is water, find the index of refraction and the speed of light in medium 2.

(a) Conceptual Reasoning:

According to Snell's law, $n_1 \sin \theta_1 = n_2 \sin \theta_2$, $n \sin \theta$ is a constant in a particular medium. Therefore, the medium with the greater angle and the sine of the angle will have a smaller index of refraction. Thus, the answer is (1) greater than because medium 2 has a smaller angle than medium 1.

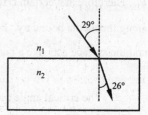

(b) Quantitative Reasoning and Solution:

Given: $n_1 = 1.33$ (water, from Table 22.1), $\theta_1 = 29°$, $\theta_2 = 26°$.

Find: n_2 and v_2.

From Snell's law, $n_1 \sin \theta_1 = n_2 \sin \theta_2$, we have $n_2 = \dfrac{n_1 \sin \theta_1}{\sin \theta_2} = \dfrac{(1.00) \sin 40°}{\sin 26°} = 1.47$.

As expected, the index of medium 2 is greater than the index of medium 1 (1.47 > 1.33).

Also, from $n = \dfrac{c}{v}$, the speed of light in medium 2 is $v_2 = \dfrac{c}{n_2} = \dfrac{3.00 \times 10^8 \text{ m/s}}{1.47} = 2.04 \times 10^8 \text{ m/s}$.

Example 22.4   A light ray from a He-Ne laser has a wavelength of 632.8 nm and travels from air to crown glass.

(a) What is the frequency of the light in air?

(b) What is the frequency of the light in crown glass?

(c) What is the wavelength of light in crown glass?

Solution: Given: $n_2 = 1.52$ (crown glass, from Table 22.1)

$n_1 = 1.00$ (air), $\lambda = 632.8$ nm $= 632.8 \times 10^{-9}$ m.

Find: (a) f (b) f_m (c) λ_m.

(a) From $c = \lambda f$, we have $f = \dfrac{c}{\lambda} = \dfrac{3.00 \times 10^8 \text{ m/s}}{632.8 \times 10^{-9} \text{ m}} = 4.74 \times 10^{14}$ Hz.

(b) The frequency is a constant (same for all media), so $f_m = f = 4.74 \times 10^{14}$ Hz.

(c) From $n = \dfrac{\lambda}{\lambda_m}$, $\lambda_m = \dfrac{\lambda}{n} = \dfrac{632.8 \text{ nm}}{1.52} = 416$ nm.

4.   Total Internal Reflection and Fiber Optics (Section 22.4)

When a ray goes from a medium of greater index of refraction to a medium of lesser index of refraction ($n_1 > n_2$), $\theta_1 < \theta_2$, according to Snell's law. When θ_1 increases, θ_2 will also increase but it will always be greater than θ_1. Therefore at a certain **critical angle** (θ_c) of incidence, the angle of refraction will be 90°, and the refracted ray is along the media boundary. For any angle of incidence $\theta_1 > \theta_c$, **total internal reflection** (no refracted light) occurs, and the surface acts like a mirror.

The critical angle can be calculated from Snell's law in terms of the indices of refraction of the two media, $\sin \theta_c = n_2/n_1$ (for $n_1 > n_2$). If the second medium is air, then $\sin \theta_c = 1/n$. **Fiber optics** uses the principle of total internal reflection. Signals can travel a long distance without losing much intensity due to the lack of refraction.

Note: Total internal reflection occurs only if the first medium has a greater index of refraction than the second medium *and* if the angle of incidence exceeds the critical angle.

Example 22.5   A diver is 1.5 m beneath the surface of a still pond of water. At what angle must the diver shine a beam of light toward the surface in order for a person on a distant bank to see it?

Solution: Given: $\theta_2 = 90°$, $n_1 = 1.00$ (air), $n_1 = 1.33$ (water).

                Find: θ_1.

When the refracted light is along the boundary, the angle in water is equal to the critical angle.

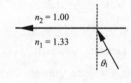

$$\theta_1 = \theta_c = \sin^{-1} \frac{n_2}{n_1} = \sin^{-1} \frac{1.00}{1.33} = \sin^{-1} 0.752 = 49°.$$

Or from Snell's law, $n_1 \sin \theta_1 = n_2 \sin \theta_2$, we have

$$\sin \theta_1 = \frac{n_2 \sin \theta_2}{n_1} = \frac{(1.00) \sin 90°}{1.33} = 0.752.$$

Thus, $\theta_1 = \sin^{-1} (0.752) = 49°$.

For the light to reach the person on the distant bank, θ_1 has to be at or slightly below 49° so there is no total internal reflection. If θ_1 exceeds 49°, then total internal reflection occurs and light will not reach the person on the distant bank. Since light is reversible, the person on the distant bank will not see the diver either.

Can you explain why some fish-preying birds stay very low before they try to catch fish?

Example 22.6   A 45°–45° prism is a wedge-shaped object in which the two acute angles are 45°. This type of prism is very useful for changing the direction of light rays by 90° in optical devices such as binoculars. There is no refracted light into air so all the light is reflected. (There is no loss of light in the direction change.) If a light ray is traveling through a glass prism according to the diagram shown, what is the minimum index of refraction of the glass?

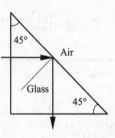

Solution: Given: $n_2 = 1.00$ (air), $\theta_c = 45°$.   Find: n_1 (minimum).

      There is no refracted ray beyond the glass–air boundary, so the light must be internally reflected. For total internal reflection to occur, the index of reflection n_1 must be greater than that of air. The minimum index of reflection corresponds to an angle of incidence of 45°, the critical angle.

$$\sin \theta_c = \frac{n_2}{n_1}, \text{ so the minimum index of refraction of the glass is}$$

$$n_1 = \frac{n_2}{\sin \theta_c} = \frac{1.00}{\sin 45°} = 1.41.$$

---

5.   Dispersion (Section 22.5)

Dispersion is the separation of multi-wavelength light into its component wavelengths when the light is refracted. This phenomenon is due to the fact that different wavelengths have slightly different speeds in a particular medium and therefore slightly different indices of refraction. In most materials (so-called normal dispersion), longer wavelengths have smaller indices of refraction. (Red light has a smaller index of refraction than blue light, for example.) According to Snell's law, different colors or wavelengths will have different angles of refraction and are therefore separated. That is the reason why we often see rainbow colors when light from the Sun enters a room through window glasses. The light from the Sun is considered white light because it has all the colors or wavelengths.

A rainbow is produced by refraction, dispersion, and total internal reflection within water droplets.

Example 22.7   A beam of white light in air strikes a piece of glass at a 70° angle (measured from the normal). A red light of wavelength 680 nm and a blue light of wavelength 430 nm emerge from the boundary after being dispersed. The index of refraction for the red light is 1.4505, and the index of refraction for the blue light is 1.4693.

(a) Which color of light is refracted more?

(b) What is the angle of refraction for each color?

(c) What is the angular separation between the two colors?

Solution: Given: $n_1 = 1.0000$ (air), $n_{2r} = 1.4505$, $n_{2b} = 1.4693$, $\theta_1 = 70°$.

Find: (b) θ_{2r} and θ_{2b} (c) $\Delta\theta$.

(a) Because the index of refraction of the blue light is greater, its angle of refraction is smaller (bent more toward the normal), and therefore it is refracted more. (Its direction is more deviated from the original direction in air.)

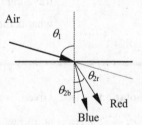

(b) We use Snell's law, $n_1 \sin \theta_1 = n_2 \sin \theta_2$.

For red: $\sin \theta_{2r} = \dfrac{n_1 \sin \theta_1}{n_{2r}} = \dfrac{(1.0000) \sin 70°}{1.4505} = 0.64784.$

So $\theta_{2r} = \sin^{-1} (0.64784) = 40.379°$;

For blue: $\sin \theta_{2b} = \dfrac{(1.0000) \sin 70°}{1.4693} = 0.63955.$

So $\theta_{2b} = \sin^{-1} (0.63955) = 39.758°$.

(c) The angular separation between the two colors is then

$\Delta\theta = \theta_{2r} - \theta_{2b} = 40.379° - 39.758° = 0.621°.$

III. Mathematical Summary

Law of Reflection	$\theta_i = \theta_r$ (22.1)	Relates the angles of incidence and reflection.
Index of Refraction	$n = \dfrac{c}{v} = \dfrac{\lambda}{\lambda_m}$ (22.3, 22.4)	Defines the index of refraction of a material.
Snell's Law	$\dfrac{\sin \theta_1}{\sin \theta_2} = \dfrac{v_1}{v_2}$ (22.2) or $n_1 \sin \theta_1 = n_2 \sin \theta_2$ (22.5)	Relates the angles of incidence and refraction, and the speeds of light in the media (or indices of refraction).
Critical Angle For Total Internal Reflection	$\sin \theta_c = \dfrac{n_2}{n_1}$, (22.6) where $n_1 > n_2$	Computes the critical angle between two materials for total internal reflection.

IV. Solutions of Selected Exercises and Paired Exercises

10. (a) If the angle of incidence is β, the angle of reflection is also β.

    So the angle formed by the left mirror and the reflecting light off the

    left mirror is $90° - \beta$.

    Then the angle between the right mirror and the light incident on the

    right mirror is $180° - [(90° - \beta) + \alpha] = 90° + \beta - \alpha$.

    Therefore the angle of incidence on the right mirror is

    $90° - [90° + \beta - \alpha] = \alpha - \beta$.   The angle of reflection off the right

    mirror is also $\alpha - \beta$ so the answer is $\boxed{(4)\ \alpha - \beta}$.

    (b) For $\alpha = 60°$ and $\beta = 40°$, the angle of reflection off the right mirror is $\alpha - \beta = 60° - 40° = \boxed{20°}$.

14. According to the law of reflection,

    $\beta = 180° - [\alpha + (90° - \theta_{i_1})] = 90° - \alpha + \theta_{i_1}$.

    So the angle of reflection from the second mirror is

    $\theta_{r_2} = 90° - \beta = \alpha - \theta_{i_1}$.

    (a) $\theta_{r_2} = 70° - 35° = \boxed{35°}$.

    (b) $\theta_{r_2} = 115° - 60° = \boxed{55°}$.

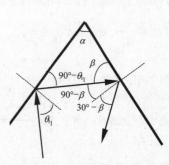

22.   Initially, light has to travel along a straight line to reach our eyes, and the top of
the container blocks the light from the bottom of the coin. When water is
added, the light coming out of water is bent into the air with a larger angle of
refraction, so it reaches our eyes.

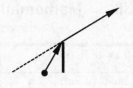

23.   The laser beam has a better chance to hit the fish. The fish appears to
the hunter at a location different from its true location due to refraction.
The laser beam obeys the same law of refraction and retraces the light the
hunter sees from the fish. The arrow goes into the water in a near-straight
line path and thus passes above the fish.

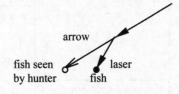

26.   (a) According to Snell's law, the angle of refraction will be (3) less than the angle of incidence because
water has a higher index of refraction.

(b) From $n_1 \sin \theta_1 = n_2 \sin \theta_2$, $\sin \theta_2 = \dfrac{n_1 \sin \theta_1}{n_2} = \dfrac{(1)\, \sin 60°}{1.33} = 0.651$.

So $\theta_2 = \sin^{-1}(0.651) = \boxed{41°}$.

34.   (a) 4/3 is greater than 5/4. According to the definition of the index of refraction, $n = \dfrac{c}{v}$, the speed of light
in material A is (3) less than because its index of refraction is higher.

(a) $\dfrac{v_A}{v_B} = \dfrac{c/n_A}{c/n_B} = \dfrac{n_B}{n_A} = \dfrac{5/4}{4/3} = \boxed{\dfrac{15}{16}}$.

37.   (a) This is caused by refraction of light in the water-air interface. The angle of refraction in air is greater
than the angle of incidence in water so the object immersed in water appears closer to the surface.

(b) From the figure, the distance a is common to both d and d', and using trigonometry

$\tan \theta_1 = \dfrac{a}{d'}$ and $\tan \theta_2 = \dfrac{a}{d}$.

Combining these two equations to form a ratio: $\dfrac{d'}{d} = \dfrac{\tan \theta_2}{\tan \theta_1}$ or $d' = \dfrac{\tan \theta_2}{\tan \theta_1}\, d$.

If $\theta < 15°$, $\tan \theta \approx \sin \theta$. So $\dfrac{d'}{d} = \dfrac{\tan \theta_2}{\tan \theta_1} \approx \dfrac{\sin \theta_2}{\sin \theta_1} = \dfrac{1}{n}$ (Snell's law).

Therefore $d' \approx \dfrac{d}{n}$.

42. The setting is at zero altitude or 90° from the normal above water, so the angle incidence in the water should be equal to the critical angle of

$$\theta_c = \sin^{-1} \frac{n_2}{n_1} = \sin^{-1} \frac{1}{1.33} = \sin^{-1}(0.752) = 48.8°.$$

Then the angle to the surface is $90° - 48.8° = \boxed{41.2°}$.

44. (a) This arrangement depends on $\boxed{\text{(3) the indices of refraction of both}}$, because $\theta_c \geq \sin^{-1} \frac{n_2}{n_1}$.

(b) Air: $\theta_c \geq \sin^{-1} \frac{n_2}{n_1}$, so $n_1 \geq \frac{n_2}{\sin \theta_c} = \frac{1}{\sin 45°} = \boxed{1.41}$.

Water: $n_1 \geq \frac{n_2}{\sin \theta_c} = \frac{1.33}{\sin 45°} = \boxed{1.88}$.

51. (a) The cause of this is due to $\boxed{\text{(3) total internal reflection}}$.

(b) At the plastic-air interface, the critical angle is

$$\theta_c = \sin^{-1} \frac{n_2}{n_1} = \sin^{-1} \frac{1}{1.60} = \sin^{-1}(0.625) = \boxed{38.7°} < 45°.$$

There is total internal reflection and the answer is $\boxed{\text{no}}$.

(c) At the plastic-liquid interface, $\theta_c = \sin^{-1} \frac{1.20}{1.60} = \sin^{-1}(0.75) = 49°$.

At the liquid-air interface, $\theta_c = \sin^{-1} \frac{1}{1.20} = \sin^{-1}(0.833) = 56°$.

The angle of refraction inside the liquid is $\sin \theta_2 = \frac{n_1 \sin \theta_1}{n_2} = \frac{(1.60) \sin 45°}{1.20} = 0.942$,

so $\theta_2 = \sin^{-1}(0.942) = 71°$.

Therefore, the angle of incidence inside the liquid (71°) is greater than the critical angle (56°); light is $\boxed{\text{still not transmitted}}$ because total internal reflection occurs at the liquid-air interface.

62. $\boxed{\text{Color B}}$ has a longer wavelength. For normal dispersion, shorter wavelength bends more than longer wavelength after refracting through a prism.

64. (a) $\boxed{\text{Blue}}$ will experience more refraction because its index of refraction differs more than for red, compared with the index of refraction of air. According to Snell's law, blue will have a smaller angle of refraction or deviates more from the angle of incidence.

(b) $n_1 \sin \theta_1 = n_R \sin \theta_R = n_B \sin \theta_B$, so $\sin \theta_R = \dfrac{n_1 \sin \theta_1}{n_R} = \dfrac{(1) \sin 37°}{1.515} = 0.3972$.

So $\theta_R = \sin^{-1} (0.3972) = 23.406°$.

$\sin \theta_B = \dfrac{(1) \sin 37°}{1.523} = 0.3952$. so $\theta_B = \sin^{-1} (0.3952) = 23.275°$.

Hence $\Delta \theta = \theta_R - \theta_B = 23.406° - 23.275° = \boxed{0.131°}$.

69.   (a) $n_A \sin \theta_A = n_B \sin \theta_B$, so $\sin \theta_B = \dfrac{n_A \sin \theta_A}{n_B} = \dfrac{n_A}{n_B} \sin \theta_A = (1.5) \sin 30° = 0.75$.

Therefore $\theta_B = \sin^{-1} (0.75) = \boxed{49°}$.

(b) Also from $n = \dfrac{c}{v}$, we have $v = \dfrac{c}{n}$. So $\dfrac{v_B}{v_A} = \dfrac{n_A}{n_B} = \boxed{1.5}$.

(c) Frequency does not change, so $\dfrac{f_B}{f_A} = \boxed{1}$.

(d) $\theta_c = \sin^{-1} \dfrac{n_B}{n_A} = \sin^{-1} (1/1.5) = \sin^{-1} 0.667 = \boxed{42°}$.

71.   (a) $\theta_c = \sin^{-1} \dfrac{n_2}{n_1}$, where $n_2 = 1$ (air). Red light will have a smaller index of refraction because its critical

angle is greater. Also because $v = \dfrac{c}{n}$, red light will have a higher speed of light than blue light. For the

same time interval red light will travel $\boxed{\text{(1) more than}}$ 1.000 m.

(b) $n_R = \dfrac{1}{\sin (\theta_c)_R} = \dfrac{1}{\sin 41.11°} = 1.521$. $n_B = \dfrac{1}{\sin (\theta_c)_B} = \dfrac{1}{\sin 41.04°} = 1.523$.

$\Delta d = \Delta v t = (v_R - v_B) t = v_B t \left(\dfrac{v_R}{v_B} - 1 \right) = v_B t \left(\dfrac{n_B}{n_A} - 1 \right)$.

Since $v_B t = 1.000$ m, $\Delta d = \left(\dfrac{1.523}{1.521} - 1 \right)(1.000 \text{ m}) = 0.0013 \text{ m} = \boxed{1.3 \text{ mm}}$.

73.   If the angle of reflection is twice the angle of refraction, the angle of incidence is also twice the angle of refraction, $\theta_1 = 2\theta_2$.

$n_1 \sin \theta_1 = n_2 \sin \theta_2$, so $n_2 = \dfrac{n_1 \sin \theta_1}{\sin \theta_2} = \dfrac{(1.00) \sin \theta_1}{\sin \theta_2} = \dfrac{\sin 2\theta_2}{\sin \theta_2} = \dfrac{2 \sin \theta_2 \cos \theta_2}{\sin \theta_2} = 2 \cos \theta_2$.

The minimum value for θ_2 is zero, so the maximum n_2 is 2.00.

Since the maximum for θ_1 is 90° or $\theta_1 = 2\theta_2 = 90°$, the maximum $\theta_2 = 45°$.

Therefore the minimum $n_2 = 2 \cos 45° = 1.41$. Thus the range is $\boxed{\text{1.41 to 2.00}}$.

V. Practice Quiz

1. If the speed of light in a material is 2.13×10^8 m/s, what is its index of refraction?

    (a) 0.710   (b) 1.07   (c) 1.41   (d) 2.13   (e) 5.13

2. A light ray in air is incident on an air-glass interface at an angle of 45° and is refracted in the glass at an angle of 27° with the normal. What is the index of refraction of the glass?

    (a) 0.642   (b) 1.16   (c) 1.41   (d) 1.56   (e) 2.20

3. An optical fiber is made of clear plastic with index of refraction of $n = 1.50$. What is the minimum angle of incidence so total internal reflection can occur?

    (a) 23.4°   (b) 32.9°   (c) 38.3°   (d) 40.3°   (e) 41.8°

4. A certain kind of glass has an index of refraction of 1.65 for blue light and an index of refraction of 1.61 for red light. If a beam of white light (containing all colors) is incident at an angle of 30°, what is the angle between the refracted red and blue light?

    (a) 0.22°   (b) 0.35°   (c) 0.45°   (d) 1.90°   (e) 1.81°

5. A ray of white light, incident on a glass prism, is dispersed into its various color components. Which one of the following colors experiences the least refraction?

    (a) orange   (b) yellow   (c) red   (d) blue   (e) green

6. Which one of the following describes what will generally happen to a light ray incident on a glass-air boundary?

    (a) total reflection   (b) total refraction   (c) partial reflection, partial refraction

    (d) either (a) or (c)   (e) either (b) or (c)

7. Light enters water from air. The angle of refraction will be

    (a) greater than or equal to the angle of incidence.   (b) less than or equal to the angle of incidence.

    (c) equal to the angle of incidence.   (d) greater than the angle of incidence.

    (e) less than the angle of incidence.

8. Dispersion can be observed in

    (a) reflection.   (b) refraction.   (c) total internal reflection.

    (d) all the preceding.   (d) none of the preceding.

9. A fiber-optic cable ($n = 1.50$) is submerged in water. What is the critical angle for light to stay inside the cable?

(a) 27.6°   (b) 41.8°   (c) 45.0°   (d) 62.5°   (e) 83.1°

10. An oil film ($n = 1.47$) floats on a water ($n = 1.33$) surface. If a ray of light is incident on the air-oil boundary at an angle of 37° to the normal, what is the angle of refraction at the oil-water boundary?

(a) 17.9°   (b) 24.2°   (c) 26.9°   (d) 33.0°   (e) 37.0°

11. The angle of incidence of a light ray entering water from air is 30°. What is the angle between the reflected ray and the refracted ray?

(a) 22°   (b) 52°   (c) 60°   (d) 68°   (e) 128°

12. A monochromatic light source emits a wavelength of 633 nm in air. When the light passes through a liquid, its wavelength decreases to 487 nm. What is the index of refraction of the liquid?

(a) 0.769   (b) 1.30   (c) 1.41   (d) 1.62   (e) 2.11

Answers to Practice Quiz:

1. c 2. d 3. e 4. c 5. c 6. d 7. e 8. b 9. d 10. c 11. e 12. b

CHAPTER 23

I.　Chapter Objectives

Upon completion of this chapter, you should be able to:

1. understand how images are formed, and describe the characteristics of plane mirrors.

2. distinguish between converging and diverging spherical mirrors, describe images and their characteristics, and determine image characteristics from ray diagrams and the spherical-mirror equation.

3. distinguish between converging and diverging lenses, describe their images and their characteristics, and find image locations and characteristics by using ray diagrams and the thin-lens equation.

4. describe the lens maker's equation, explain how it differs from the thin-lens equation, and understand lens "power" in diopters.

*5. describe some common lens aberrations, and explain how they can be reduced or corrected.

II.　Chapter Summary and Discussion

1.　Plane Mirrors (Section 23.1)

The images of objects formed by optical systems (mirrors and/or lenses) can be either real or virtual. A **real image** is one formed by light rays that converge at and pass through the image location, and can be projected onto or formed on a screen. A **virtual image** is one for which light rays *appear* to emanate from the image but do not actually do so. Virtual images cannot be projected onto or formed on screens.

A **plane mirror** is a mirror with a flat surface. A mirror forms an image based on the law of reflection. The characteristics of the images formed by a plane mirror are always virtual, upright, and unmagnified ($M = +1$). Also, the image formed by a plane mirror appears to be at a distance behind the mirror that is equal to the distance of the object in front of the mirror ($d_o = d_i$) and has right-left or front-back reversal.

Example 23.1　A curved arrow is placed in front of a plane mirror as shown. Sketch the image of the arrow formed by the plane mirror.

Solution:

The image is virtual, upright, and unmagnified. The object distance (distance from object to mirror) is equal to the image distance (distance from image to mirror) for images formed by a plane mirror. By sketching the image of each individual point on the object, we obtain the image.

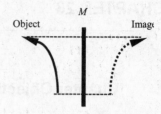

2. Spherical Mirrors (Section 23.2)

A **spherical mirror** is a section of a sphere. Either the outside (convex) surface or the inside (concave) surface of the spherical section may be the reflecting surface. A **concave mirror** is called a **converging mirror**, and a **convex mirror** is called a **diverging mirror**. These terms refer to the reflections of rays parallel to a mirror's *optical axis*, which is a line through the center of the spherical mirror that intersects the mirror at the *vertex* of the spherical section.

The **center of curvature** (C) of a spherical mirror is the point on the optic axis that corresponds to the center of the sphere of which the mirror forms a section. The **radius of curvature** (R) is the distance from the vertex to the center of curvature. The **focal point** (F) of a spherical mirror is the point at which parallel rays converge or appear to diverge. The **focal length** (f) of a spherical mirror is the distance from the focal point to the vertex of the spherical section. It is equal to half of the radius of curvature, $f = R/2$.

The images formed by spherical mirrors can be studied from geometry (**ray diagrams**). The ray diagram is as important as the free-body diagram in the application of Newton's laws. Three important rays are used to determine the images: a **parallel ray** is a ray that is incident along a path parallel to the optic axis and is reflected through (or appears to go through) the focal point F ; a **chief (radial) ray** is a ray that is incident through (or appears to go through) the center of curvature C and is reflected back along its incident path though C; a **focal ray** is a ray that passes through (or appears to go through) the focal point and is reflected parallel to the optic axis.

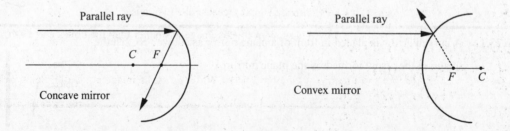

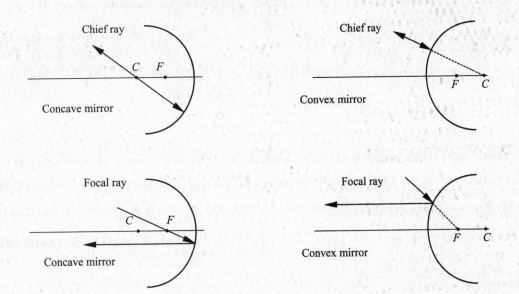

Chief ray

C F

Concave mirror

Chief ray

F C

Convex mirror

Focal ray

C F

Concave mirror

Focal ray

F C

Convex mirror

The images of objects can be *upright*, or *inverted*, *magnified*, *unmagnified*, or *reduced in size*. All these characteristics can be determined from the **lateral magnification**, $M = \dfrac{\text{image height}}{\text{object height}} = \dfrac{h_i}{h_o}$. For real objects (these are the only ones we will deal with), if M is negative, the image is real and inverted; if M is positive, the image is virtual and upright; if $|M| > 1$, the image is magnified; if $|M| = 1$, the image is unmagnified; and if $|M| < 1$, the image is reduced. These image characteristics are very important in geometrical optics.

Example 23.2 An object is placed 30 cm in front of a concave mirror of radius 20 cm. Determine the characteristics of the image with a ray diagram.

Solution:

The focal length of the mirror is $f = \dfrac{R}{2} = \dfrac{20 \text{ cm}}{2} = 10$ cm.

We use the parallel ray (parallel in, through focal point out) and chief ray (through center of curvature in, through center of curvature out). These two rays originate from the tip of the object and cross after being reflected by the mirror. The point where they cross is the image point, so the image is real (because the two rays converged), inverted (the image is upside down), and reduced (the image is smaller than the object), as determined from the ray diagram.

Try to draw the focal ray from the tip of the object here. Will it cross the image point after being reflected?

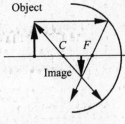

Object

C F

Image

The images formed by spherical mirrors can also be studied from mathematical equations. The **spherical-mirror equation** is given by $\dfrac{1}{d_o} + \dfrac{1}{d_i} = \dfrac{1}{f}$, or $d_i = \dfrac{d_o f}{d_o - f}$, where d_o is the object distance (from the object to the vertex), d_i is the image distance (from the image to the vertex), and f is the focal length. The lateral magnification can be written in terms of the object and image distances as $M = \dfrac{h_i}{h_o} = -\dfrac{d_i}{d_o}$.

Note: d_i and f should be used as algebraic quantities, that is, they have signs (positive or negative). For a real object (positive d_o), if a mirror is concave (converging), f is positive; if a mirror is convex (diverging), f is negative; if the image is real (formed on the same side of the mirror as the object), d_i is positive; if the image is virtual (formed behind the mirror), d_i is negative. Also, $\dfrac{1}{d_o}$, $\dfrac{1}{d_i}$, or $\dfrac{1}{f}$ are usually calculated first, so do not forget to take the inverse to get d_i, d_o, or f.

Example 23.3   Find the location and describe the characteristics of the image formed by a concave mirror of radius 20 cm if the object distance is   (a) 30 cm   (b) 20 cm   (c) 15 cm   (d) 5.0 cm.

Solution: Given: $R = 20$ cm,   (a) $d_o = 30$ cm   (b) $d_o = 20$ cm   (c) $d_o = 15$ cm   (d) $d_o = 5.0$ cm.
           Find: d_i and characteristics in (a), (b), (c), and (d).

The focal length of the mirror is $f = \dfrac{R}{2} = \dfrac{20 \text{ cm}}{2} = 10$ cm.

(a) The object distance is greater than two focal lengths (or the radius of curvature), $d_o > 2f = R$.

From $\dfrac{1}{d_o} + \dfrac{1}{d_i} = \dfrac{1}{f}$, we have $\dfrac{1}{30 \text{ cm}} + \dfrac{1}{d_i} = \dfrac{1}{10 \text{ cm}}$.

So $\dfrac{1}{d_i} = \dfrac{1}{10 \text{ cm}} - \dfrac{1}{30 \text{ cm}} = \dfrac{3}{30 \text{ cm}} - \dfrac{1}{30 \text{ cm}} = \dfrac{2}{30 \text{ cm}}$.

Therefore, $d_i = \dfrac{30 \text{ cm}}{2} = 15$ cm.

The lateral magnification is equal to $M = -\dfrac{d_i}{d_o} = -\dfrac{15 \text{ cm}}{30 \text{ cm}} = -\dfrac{1}{2}$.

The image is *real* (d_i is positive), *inverted* (M is negative), and *reduced* ($|M| < 1$) for $d_o > 2f = R$.

(b) The object distance is equal to two focal lengths (or the radius of curvature), $d_o = 2f = R$.

$\dfrac{1}{d_i} = \dfrac{1}{10 \text{ cm}} - \dfrac{1}{20 \text{ cm}} = \dfrac{2}{20 \text{ cm}} - \dfrac{1}{20 \text{ cm}} = \dfrac{1}{20 \text{ cm}}$.

So $d_i = 20$ cm and $M = -\dfrac{20 \text{ cm}}{20 \text{ cm}} = -1$.

The image is *real* (d_i is positive), *inverted* (M is negative), and *unmagnified* ($|M| = 1$) for $d_o = 2f = R$.

(c) The object distance is greater than one focal length and smaller than two focal lengths, $f < d_o < 2f = R$.

$$\frac{1}{d_i} = \frac{1}{10 \text{ cm}} - \frac{1}{15 \text{ cm}} = \frac{3}{30 \text{ cm}} - \frac{2}{30 \text{ cm}} = , \text{ so } d_i = 30 \text{ cm, and } M = -\frac{30 \text{ cm}}{15 \text{ cm}} = -2.$$

The image is *real* (d_i is positive), *inverted* (M is negative), and *magnified* ($|M| > 2$) for $f < d_o < 2f = R$.

(d) The object distance is smaller than one focal length, $d_o < f = R/2$.

Here let's use the alternative form of the spherical-mirror equation:

$$d_i = \frac{d_o f}{d_o - f} = \frac{(5.0 \text{ cm})(10 \text{ cm})}{5.0 \text{ cm} - 10 \text{ cm}} = -10 \text{ cm and } M = -\frac{-10 \text{ cm}}{5.0 \text{ cm}} = +2.$$

The image is *virtual* (d_i is negative), *upright* (M is positive), and *magnified* ($|M| > 1$) for $d_o < f = R/2$.

Hence, a concave mirror can form a variety of images depending on the object distance. The images formed here could be real or virtual, magnified or reduced, erect or inverted.

Integrated Example 23.4

When a person's face is 40 cm in front of a cosmetic mirror (concave mirror), the erect image is three times the size of the object. (a) The distance from the image to the mirror is (1) greater than, (2) the same as, or (3) less than the distance from the object to the mirror. Explain. (b) What are the image distance and the focal length of this mirror?

(a) Conceptual Reasoning:

Because the image is erect and three times the size, the lateral magnification is equal to +3.0 ($M = +3.0$).

Since the lateral magnification is $M = -\dfrac{d_i}{d_o}$, the magnitude of d_i is greater than the magnitude of d_o so the answer is (1) greater than.

(b) Quantitative Reasoning and Solution:

Given: $d_o = 40 \text{ cm}$, $M = +3.0$ (erect image). Find: d_i and f.

From the lateral magnification $M = -\dfrac{d_i}{d_o}$, we have $d_i = -M d_o = -(+3.0)(40 \text{ cm}) = -120 \text{ cm}$.

As expected, the distance from the image to the mirror (120 cm) is greater than the distance from the object to the mirror (40 cm).

Using the spherical-mirror equation, $\dfrac{1}{d_o} + \dfrac{1}{d_i} = \dfrac{1}{f}$, $\dfrac{1}{f} = \dfrac{1}{40 \text{ cm}} + \dfrac{1}{-120 \text{ cm}} = \dfrac{1}{60 \text{ cm}}$.

So $f = 60 \text{ cm}$.

Example 23.5   What are the image characteristics of an object 30 cm from a convex mirror of radius 120 cm?

Solution: Given: $d_o = 30$ cm, $R = -120$ cm (convex mirror).   Find: image characteristics.

For a convex mirror, the radius and focal length are negative: $f = \dfrac{R}{2} = \dfrac{-120 \text{ cm}}{2} = -60$ cm.

From the spherical-mirror equation, we have $\dfrac{1}{30 \text{ cm}} + \dfrac{1}{d_i} = \dfrac{1}{-60 \text{ cm}}$.

So $\dfrac{1}{d_i} = \dfrac{1}{-60 \text{ cm}} - \dfrac{1}{30 \text{ cm}} = -\dfrac{1}{60 \text{ cm}} - \dfrac{2}{60 \text{ cm}} = -\dfrac{3}{60 \text{ cm}} = -\dfrac{1}{20 \text{ cm}}$.

Therefore, $d_i = -20$ cm. Because d_i is negative, the image is virtual.

The lateral magnification is equal to $M = -\dfrac{d_i}{d_o} = -\dfrac{-20 \text{ cm}}{30 \text{ cm}} = +2/3$.

Because $M > 0$ and $|M| < 1$, the image is upright and reduced.

3.   Lenses (Section 23.3)

A lens forms an image based on the law of refraction (Snell's law). A spherical **biconvex lens** is a **converging lens**, and a spherical **biconcave lens** is a **diverging lens**. For these lenses, the focal length is *not* equal to half the radius of curvature of the surface. Only thin lenses are studied here to ignore the lateral displacement of the ray after it is refracted from the two lens surfaces.

The images formed by thin lenses can also be studied from geometry (**ray diagrams**). The general rules for drawing ray diagrams for lenses are similar to those for spherical mirrors, with some modifications. Three important rays are used to determine the images: a **parallel ray** is a ray that is incident along a path parallel to the optic axis and is refracted through (or appears to go through) the focal point F; a **chief (central) ray** is a ray that is incident through (or appears to go through) the center of the lens and is refracted undeviated; a **focal ray** is a ray that passes through (or appears to go through) the focal point and is refracted parallel to the optic axis.

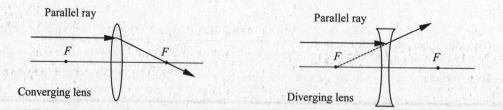

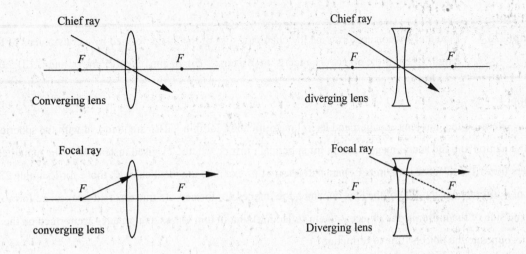

Converging lens

diverging lens

Focal ray

Focal ray

converging lens

Diverging lens

Example 23.6 An object is placed 5.0 cm in front of a thin lens of focal length 15 cm. Determine the characteristics of the image with a ray diagram.

Solution:

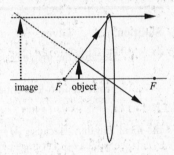

We use the chief ray (through center of lens in, undeviated out) and the focal ray (through focal point in, parallel out). These two rays originate from the tip of the object and *appear* to cross after being refracted by the lens. The point where they appear to cross is the image point, so the image is virtual (because the two rays do not cross), upright (the image is right-side up), and magnified (the image is larger than the object), as determined from the ray diagram.

Try to draw the parallel ray from the tip of the object here. Will it cross the image point after being refracted?

The images formed by thin lenses can also be studied from mathematical equations. The **thin-lens equation** is identical in form with the spherical-mirror equation, $\frac{1}{d_o} + \frac{1}{d_i} = \frac{1}{f}$, or $d_i = \frac{d_o f}{d_o - f}$, where d_o is the object distance (from the object to the center of the lens), d_i is the image distance (from the image to the center of the lens), and f is the focal length. The lateral magnification is also defined as $M = \frac{h_i}{h_o} = -\frac{d_i}{d_o}$.

Note: d_i and f should be used as algebraic quantities, that is, they have signs (positive or negative). For real objects (positive d_o), if a lens is convex (converging), f is positive; if a lens is concave (diverging), f is negative; if the image is real (formed on the side of the lens opposite the object), d_i is positive; if the image is virtual (formed on the same side as the object), d_i is negative. Also, a concave mirror is converging, but a concave lens is diverging; a convex mirror is diverging, but a convex lens is converging.

Example 23.7   Find the location and describe the characteristics of the image formed by a convex lens of focal length 10 cm if the object distance is (a) 30 cm,   (b) 20 cm,   (c) 15 cm,   and   (d) 5.0 cm.

Solution:

Because the thin-lens equation and lateral magnification for thin lenses are identical with the spherical-mirror equation and the lateral magnification for spherical mirrors, where $2f$ is used instead of R, the numerical answers for image distances and image characteristics in this example are identical with those for Example 23.3. The only difference is the side of the lens on which the images are formed. In mirrors, real images are formed on the same side of the mirror as the objects (due to reflection), but in thin lenses, real images are formed on the side of the lens opposite the objects (due to refraction).

Example 23.8   A biology student uses a convex thin-lens to examine a small worm that is 4.0 cm in front of the lens. The image is upright and twice the size of the worm. What is the focal length of the lens?

Solution: Given: $d_o = 4.0$ cm, $M = +2.0$ (upright).   Find: f.

Because the image is upright and twice the size, the lateral magnification is $+2.0$, $(M = +2.0)$.

From the thin-lens equation, $\dfrac{1}{d_o} + \dfrac{1}{d_i} = \dfrac{1}{f}$, we first need to know the image distance d_i before we can find

the focal length f. Because $M = -\dfrac{d_i}{d_o}$, $d_i = -Md_o = -(+2.0)(4.0 \text{ cm}) = -8.0$ cm.

Therefore, $\dfrac{1}{f} = \dfrac{1}{4.0 \text{ cm}} + \dfrac{1}{-8.0 \text{ cm}} = \dfrac{1}{8.0 \text{ cm}}$, or $f = 8.0$ cm.

Example 23.9   Determine the location and describe the characteristics of the image of an object placed at 30 cm in front of a concave (diverging lens) with a focal length of 10 cm.

Solution: Given: $f = -10$ cm (diverging), $d_o = 30$ cm.

Find: d_i and image characteristics.

The focal length of a diverging lens is negative. From the alternative form of the thin-lens equation,

$d_i = \dfrac{d_o f}{d_o - f} = \dfrac{(30 \text{ cm})(-10 \text{ cm})}{30 \text{ cm} - (-10 \text{ cm})} = -7.5$ cm. Since the image distance is negative, the image is virtual.

The lateral magnification is $M = -\dfrac{d_i}{d_o} = -\dfrac{-7.5 \text{ cm}}{30 \text{ cm}} = 0.25$, so the image is upright and reduced.

Many optical systems such as telescopes and compound microscopes use more than one lens. When two or more lenses are used in combination, the overall image produced may be determined by considering the lenses individually in sequence. That is, the image formed by the first lens is the object for the second lens, and so on. If the first lens produces an image in front of the second lens, the image is treated as a real object for the second lens. If, however, the lenses are close enough together that the image from the first lens is not formed in front of the second lens, then a modification must be made in the sign convention. In this case, the image from the first lens is treated as a *virtual* object for the second lens, and the object distance for it is taken to be *negative* in the lens equation. The total magnification (M_{total}) of a compound lens system is the product of the lateral magnifications of all the component lenses. For a two-lens system, $M_{total} = M_1 \times M_2$.

4. Lens Maker's Equation (Section 23.4)

The **lens maker's equation** is a general equation for determining the focal length of a spherical lens with sides having different (or equal) radii of curvature. In general, the focal length of a lens in air ($n_{air} = 1$) can be calculated from $\dfrac{1}{f} = (n-1)\left(\dfrac{1}{R_1} + \dfrac{1}{R_2}\right)$, where n is the index of refraction of the lens material, and R_1 and R_2 are the radii of the first (front) and second (back) surfaces. The sign convention for the radii are as follows: if the center of curvature C is on the side of the lens from which light emerges or on the back side of the lens, R is positive; if C is on the side of the lens on which light is incident or on the front side of the lens, R is negative. If the lens of a material with an index of refraction n is in a fluid with an index of refraction n_m, the lens maker's equation is modified to $\dfrac{1}{f} = \left(\dfrac{n}{n_m} - 1\right)\left(\dfrac{1}{R_1} + \dfrac{1}{R_2}\right)$.

The power of a lens is defined as $P = \dfrac{1}{f}$, where f is the focal length in meters. The unit of lens power is expressed in **diopters** (D) when f is in meters.

Example 23.10   A double convex lens made of glass ($n = 1.52$) has a radius of curvature of 50.0 cm on the front side and 40.0 cm on the back side. Find the power and the focal length of the lens.

Solution: Given: $n = 1.52$, $R_1 = +50.0$ cm, $R_2 = +40.0$ cm. (See sign convention on Table 23.4.)
           Find: f.

According to the sign convention, the radii for both surfaces are positive because both surfaces are convex.

From the lens maker's equation, we have

$$P = \frac{1}{f} = (n-1)\left(\frac{1}{R_1} + \frac{1}{R_2}\right) = (1.52-1)\left(\frac{1}{0.50 \text{ m}} + \frac{1}{0.40 \text{ m}}\right) = 2.34 \text{ m}^{-1} = 2.34 \text{ D}.$$

So $f = \frac{1}{P} = \frac{1}{2.34 \text{ D}} = 0.427 \text{ m} = 42.7 \text{ cm}.$

---

*5. Lens Aberrations (Section 23.5)

Rays passing through a lens may not exactly follow the particular rays in a ray diagram because of aberrations. **Spherical aberration** occurs when parallel rays passing through different regions of a lens do not converge or come together on a common image plane. (The thin-lens equation is an approximation for rays very close to the optical axis.) **Chromatic aberration** results from the fact that different wavelengths (different colors) have different indices of refraction, and so different angles of refraction, and thus different image planes for different colors. **Astigmatism** results when a cone (circular cross-section) of light from an off-axis source falls on a lens surface and forms an elliptically illuminated area, giving rise to two images. All these aberrations can be minimized, if not eliminated, by compounding lens systems. For example, some of the common 50-mm camera lenses are actually made up of seven lenses to reduce aberrations.

III. Mathematical Summary

Plane Mirror Object and Image Distances	$d_o = d_i$ (23.1)	The object distance and image distance are equal for a plane mirror.
Lateral Magnification Factor for All Mirrors and Lenses	$M = -\dfrac{d_i}{d_o}$ (23.4, 23.6)	Defines the lateral magnification as a ratio of the size of the image to the size of the object.
Focal Length of A Spherical Mirror	$f = \dfrac{R}{2}$ (23.2)	Defines the focal length of a spherical mirror in terms of its radius of curvature.
Spherical-Mirror Equation	$\dfrac{1}{d_o} + \dfrac{1}{d_i} = \dfrac{1}{f}$ or $d_i = \dfrac{d_o f}{d_o - f}$ (23.3)	Relates object distance, image distance, and focal length of a spherical mirror.
Thin-Lens Equation	$\dfrac{1}{d_o} + \dfrac{1}{d_i} = \dfrac{1}{f}$ or $d_i = \dfrac{d_o f}{d_o - f}$ (23.5)	Relates object distance, image distance, and focal length of a thin lens.

Lens Maker's Equation	$\frac{1}{f} = (n-1)\left(\frac{1}{R_1} + \frac{1}{R_2}\right)$ (23.8)	Calculates the focal length of a thin lens in air.
Lens Power in Diopters (f in meters)	$P = \frac{1}{f}$ (23.9)	Defines the power of a lens in terms of its focal length.

V.   Solutions of Selected Exercises and Paired Exercises

10.   (a) Image distance equals object distance ($d_i = d_o$).

So the distance from object to image is 40 cm + 40 cm = $\boxed{0.80 \text{ m}}$.

(b) The image has the same height as the object. So it is $\boxed{5.0 \text{ cm}}$.

(c) The image is unmagnified and upright. So the magnification is $\boxed{+1.0}$.

14.   (a) The image of the back of her head will be from $\boxed{\text{both mirrors}}$ because the image formed by one mirror

will form another image by the other image.

(b) The image formed by the wall mirror is 0.90 m + 0.90 m = 1.80 m behind the woman or

1.80 m + 0.30 m = 2.10 m behind the hand mirror. The image formed by the hand mirror is then 2.10 m in

front of the hand mirror or 2.10 m + 0.30 m = $\boxed{2.4 \text{ m}}$ in front of the woman.

23.   (a) A spoon can behave as either a concave or a convex mirror depending on which side you use for

reflection. If you use the concave side, you normally see an inverted image. If you use the convex side,

you always see an upright image.

(b) In theory, the answer is yes. If you are very close (inside the focal point) to the spoon on the concave

side, an upright image exists. However, it might be difficult for you to see the image in practice, because

your eyes might be too close to the image. Eyes cannot see things that are closer than the near point

(Chapter 25).

28. $d_o = 20 \text{ cm}$, $f = \frac{R}{2} = \frac{30 \text{ cm}}{2} = 15 \text{ cm}$. $d_i = \frac{d_o f}{d_o - f} = \frac{(20 \text{ cm})(15 \text{ cm})}{20 \text{ cm} - 15 \text{ cm}} = \boxed{60 \text{ cm}}$.

$M = -\frac{d_i}{d_o} = -\frac{60 \text{ cm}}{20 \text{ cm}} = -3.0$. So $h_i = M h_o = -3.0 \,(3.0 \text{ cm}) = -9.0 \text{ cm}$.

Therefore it is $\boxed{9.0 \text{ cm}}$ tall.

32. (a)

$d_o = 40$ cm

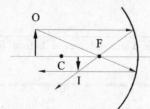

$d_o = 30$ cm

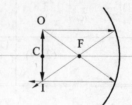

$d_o = 15$ cm

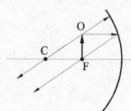

$d_o = 5.0$

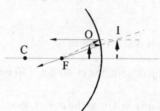

(b) $f = \dfrac{R}{2} = \dfrac{30 \text{ cm}}{2} = 15$ cm. $d_o = 40$ cm: $d_i = \dfrac{d_o f}{d_o - f} = \dfrac{(40 \text{ cm})(15 \text{ cm})}{40 \text{ cm} - 15 \text{ cm}} = \boxed{24 \text{ cm}}$.

$M = -\dfrac{d_i}{d_o} = -\dfrac{24 \text{ cm}}{40 \text{ cm}} = -0.60.$ $h_i = M h_o = -0.60\,(3.0 \text{ cm}) = -1.8 \text{ cm} = \boxed{1.8 \text{ cm, real and inverted}}$.

$d_o = 30$ cm: $d_i = \dfrac{(30 \text{ cm})(15 \text{ cm})}{30 \text{ cm} - 15 \text{ cm}} = \boxed{30 \text{ cm}}$.

$M = -\dfrac{30 \text{ cm}}{30 \text{ cm}} = -1.0.$ $h_i = -1.0\,(3.0 \text{ cm}) = -3.0 \text{ cm} = \boxed{3.0 \text{ cm, real and inverted}}$.

$d_o = 15$ cm: $d_i = \dfrac{(15 \text{ cm})(15 \text{ cm})}{15 \text{ cm} - 15 \text{ cm}} = \boxed{\infty}$. The characteristics of the image are not defined.

$d_o = 5.0$ cm: $d_i = \dfrac{(5.0 \text{ cm})(15 \text{ cm})}{5.0 \text{ cm} - 15 \text{ cm}} = \boxed{-7.5 \text{ cm}}$.

$M = -\dfrac{-7.5 \text{ cm}}{5.0 \text{ cm}} = +1.5.$ $h_i = +1.5\,(3.0 \text{ cm}) = \boxed{4.5 \text{ cm, virtual and upright}}$.

36. Since $d_o < f$, $d_i = \dfrac{d_o f}{d_o - f} < 0$. Also $M = -\dfrac{d_i}{d_o} = -\dfrac{f}{d_o - f} = \dfrac{f}{f - d_o} > +1.$

So the image is virtual (negative d_i), upright (positive M), and magnified ($|M| > 1$).

40. (a) The mirror is $\boxed{\text{convex, as the erect image is smaller than the object}}$. Only a convex mirror can form an image that is erect and reduced.

(b) $d_o = 18$ cm, $M = +\frac{1}{2}$ (it is a virtual and upright image). $d_i = -M d_o = -\frac{1}{2}(18 \text{ cm}) = -9.0$ cm.

$\dfrac{1}{f} = \dfrac{1}{d_o} + \dfrac{1}{d_i} = \dfrac{1}{18 \text{ cm}} + \dfrac{1}{-9.0 \text{ cm}} = -\dfrac{1}{18 \text{ cm}},$ so $f = \boxed{-18 \text{ cm}}$.

42. $d_o = 20$ cm, $M = +1.5$ (upright image). $d_i = -Md_o = -1.5\,(20\text{ cm}) = -30$ cm.

$$\frac{1}{f} = \frac{1}{d_o} + \frac{1}{d_i} = \frac{1}{20\text{ cm}} + \frac{1}{-30\text{ cm}} = \frac{1}{60\text{ cm}}, \text{ so } f = 60 \text{ cm.}$$

Therefore $R = 2f = 120$ cm $= \boxed{1.2 \text{ m}}$.

46. (a) The mirror is $\boxed{\text{concave, as the image is magnified}}$. Only concave mirror can form a magnified image.

(b) $d_o = 12$ cm, $M = \dfrac{9.0\text{ cm}}{3.0\text{ cm}} = +3.0$. $d_i = -Md_o = -3.0\,(12\text{ cm}) = -36$ cm.

So $\dfrac{1}{f} = \dfrac{1}{d_o} + \dfrac{1}{d_i} = \dfrac{1}{12\text{ cm}} + \dfrac{1}{-36\text{ cm}} = \dfrac{1}{18\text{ cm}}$, so $f = 18$ cm.

Therefore $R = 2f = \boxed{36 \text{ cm}}$.

51. (a) There could be $\boxed{\text{two}}$ object distances. One for a real image and another for a virtual image.

(b) $f = \dfrac{R}{2} = \dfrac{20\text{ cm}}{2} = 10$ cm, $M = \pm 2.0$, the + is for a virtual image and the − is for a real image.

$d_i = -M d_o = \mp 2.0 d_o$. $\dfrac{1}{f} = \dfrac{1}{d_o} + \dfrac{1}{d_i}$, so $\dfrac{1}{10\text{ cm}} = \dfrac{1}{d_o} + \dfrac{1}{\mp 2.0 d_o}$.

Or $\dfrac{2 \mp 1}{2 d_o} = \dfrac{1}{10\text{ cm}}$. Solving, $d_o = \dfrac{10\text{ cm}}{2}(2 \mp 1) = \boxed{5.0 \text{ cm or } 15 \text{ cm}}$.

62. $\dfrac{1}{f} = \dfrac{1}{d_o} + \dfrac{1}{d_i} = \dfrac{1}{30\text{ cm}} + \dfrac{1}{15\text{ cm}} = \dfrac{1}{10\text{ cm}}$, so $f = \boxed{10 \text{ cm}}$.

64. (a) It is seen from the ray diagram below that the image is $\boxed{\text{virtual, upright, magnified}}$.

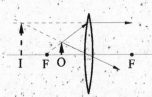

(b) $d_i = \dfrac{d_o f}{d_o - f} = \dfrac{(15\text{ cm})(22\text{ cm})}{15\text{ cm} - 22\text{ cm}} = \boxed{-47 \text{ cm}}$.

$M = -\dfrac{d_i}{d_o} = -\dfrac{-47.1\text{ cm}}{15\text{ cm}} = \boxed{+3.1}$.

70. $d_o = 4.0$ m, $M = \dfrac{h_i}{h_o} = \dfrac{-35 \text{ mm}}{1.7 \times 10^3 \text{ mm}} = -0.0206$ (inverted).

         $d_i = -Md_o = 0.0206\,(4.0 \text{ m}) = 0.0824$ m.

         $\dfrac{1}{f} = \dfrac{1}{d_o} + \dfrac{1}{d_i} = \dfrac{1}{4.0 \text{ m}} + \dfrac{1}{0.0824 \text{ m}} = 12.4 \text{ m}^{-1}$, so $f = 0.081$ m = $\boxed{8.1 \text{ cm}}$.

74. $d_o = 10$ cm, $M = +\frac{1}{5}$ (image by concave lens is always virtual). $d_i = -Md_o = -\frac{1}{5}d_o$.

         $\dfrac{1}{f} = \dfrac{1}{d_o} + \dfrac{1}{d_i} = \dfrac{1}{d_o} + \dfrac{1}{-\frac{1}{5}d_o} = \dfrac{-4}{d_o}$, so $f = -\dfrac{d_o}{4} = -\dfrac{10 \text{ cm}}{4} = \boxed{-2.5 \text{ cm}}$.

78.   (a) The lens should be $\boxed{\text{convex, as the image is magnified}}$. Only convex lens can form magnified images.

         (b) $d_o = 5.00$ cm, $M = +5.00$.   Since $M = -\dfrac{d_i}{d_o}$, $d_i = -Md_o = -(5.00)(5.00 \text{ cm}) = -25.0$ cm.

         $\dfrac{1}{f} = \dfrac{1}{d_o} + \dfrac{1}{d_i} = \dfrac{1}{5.00 \text{ cm}} + \dfrac{1}{-25.0 \text{ cm}} = \dfrac{4}{25.0 \text{ cm}}$. So $f = \dfrac{25.0 \text{ cm}}{4} = \boxed{6.25 \text{ cm}}$.

82.   For L_1, $d_{i1} = \dfrac{d_{o1}f_o}{d_{o1} - f_o} = \dfrac{(50 \text{ cm})(30 \text{ cm})}{50 \text{ cm} - 30 \text{ cm}} = 75$ cm.

         $M_1 = -\dfrac{d_{i1}}{d_{o1}} = -\dfrac{75 \text{ cm}}{50 \text{ cm}} = -1.5$. The image by L_1 is the object for L_2.

         For L_2, $d_{o2} = d - d_{i1} = 60 \text{ cm} - 75 \text{ cm} = -15$ cm, where d is the distance between the lenses. A negative object means that the "object" is on the image side.

         $d_{i2} = \dfrac{d_{o2}f_2}{d_{o2} - f_2} = \dfrac{(-15 \text{ cm})(20 \text{ cm})}{-15 \text{ cm} - 20 \text{ cm}} = \boxed{8.6 \text{ cm}}$. $M_2 = -\dfrac{d_{i2}}{d_{o2}} = -\dfrac{8.57 \text{ cm}}{-15 \text{ cm}} = 0.57$.

         So $M_{\text{total}} = M_1 M_2 = (-1.5)(0.57) = -0.86 = \boxed{0.86; \text{ real and inverted}}$.

92.   From $P = \dfrac{1}{f}$, we have $f = \dfrac{1}{P} = \dfrac{1}{-2.0 \text{ D}} = \boxed{-0.50 \text{ m}}$.

96.   (a) According to the sign convention, the signs are $\boxed{(2) +, -}$.

         (b) $\dfrac{1}{f} = (n-1)\left(\dfrac{1}{R_1} + \dfrac{1}{R_2}\right) = (1.55 - 1)\left(\dfrac{1}{0.0250 \text{ m}} + \dfrac{1}{-0.0300 \text{ m}}\right) = 3.67 \text{ m}^{-1}$.

         So $f = \dfrac{1}{3.67 \text{ /m}} = 0.272 \text{ m} = \boxed{27.2 \text{ cm}}$.

101. $f_1 = f_2 = \dfrac{1}{P} = \dfrac{1}{10\ \text{D}} = 0.10\ \text{m} = 10\ \text{cm}.$

For the first lens, $d_{i1} = \dfrac{d_{o1}f_1}{d_{o1} - f_1} = \dfrac{(60\ \text{cm})(10\ \text{cm})}{60\ \text{cm} - 10\ \text{cm}} = 12\ \text{cm}, \quad M_1 = -\dfrac{d_{i1}}{d_{o1}} = -\dfrac{12\ \text{cm}}{60\ \text{cm}} = -0.20.$

The image by the first lens is the object for the second lens.

For the second lens, $d_{o2} = d - d_{i1} = 20\ \text{cm} - 12\ \text{cm} = 8.0\ \text{cm},$ (d is the distance between the two lenses.)

$d_{i2} = \dfrac{d_{o2}f_2}{d_{o2} - f_2} = \dfrac{(8.0\ \text{cm})(10\ \text{cm})}{8.0\ \text{cm} - 10\ \text{cm}} = -40\ \text{cm},$ i.e., 40 cm on the object side from lens 2.

Therefore the position of the image relative to lens 1 is

$20\ \text{cm} - 40\ \text{cm} = -20\ \text{cm} = \boxed{20\ \text{cm on object side of first lens}}.$

$M_2 = -\dfrac{d_{i2}}{d_{o2}} = -\dfrac{-40\ \text{cm}}{8.0\ \text{cm}} = +5.0. \quad M_{\text{total}} = M_1 M_2 = (-0.20)(5.0) = \boxed{-1.0,\ \text{virtual and inverted}}.$

V.   Practice Quiz

1. Which one of the following describes the image of a concave mirror when the object distance from the mirror is greater than twice the focal length?

    (a) virtual, upright, and magnified   (b) real, inverted, and reduced

    (c) virtual, upright, and reduced   (d) real, inverted, and magnified

    (e) virtual, inverted, and magnified

2. A concave mirror with a radius of curvature of 20 cm creates a real image 30 cm from the mirror. What is the object distance?

    (a) 20 cm   (b) 15 cm   (c) 7.5 cm   (d) 5.0 cm   (e) 2.5 cm

3. A person's face is 30 cm in front of a concave shaving mirror. If the image is an upright image 1.5 times as large as the object, what is the mirror's focal length?

    (a) 12 cm   (b) 20 cm   (c) 50 cm   (d) 70 cm   (e) 90 cm

4. Which of the following describes the image of a convex mirror?

    (a) virtual, inverted, and magnified   (b) real, inverted, and reduced

    (c) virtual, upright, and reduced   (d) virtual, upright, and magnified

    (e) virtual, upright, and unmagnified

5. An object is placed at a distance of 40 cm from a thin lens. If a virtual image forms at a distance of 50 cm from the lens, on the same side as the object, what is the focal length of the lens?

    (a) 200 cm   (b) 90 cm   (c) 75 cm   (d) 45 cm   (e) 22 cm

6.   If you stand 2.5 ft in front of a plane mirror, how far away will your image be in the mirror?

      (a) 2.5 ft   (b) 5.0 ft   (c) 7.5 ft   (d) 10 ft   (e) 20 ft

7.   An object is placed at a distance of 30 cm from a thin convex lens. The lens has a focal length of 10 cm.
      What are the values, respectively, of the image distance and lateral magnification?

      (a) 65 cm, 2.0   (b) 15 cm, 2.0   (c) 25 cm, 1.0   (d) 60 cm, –0.50   (e) 15 cm, –0.50

8.   The ray diagram is an example of a

      (a) camera.   (b) overhead projector.

      (c) magnifying glass.   (d) microscope.

      (e) telescope.

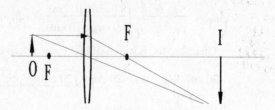

9.   What causes spherical aberration?

      (a) A ray is too far from the optic axis.

      (b) A different color has the same index of refraction.

      (c) A circular cone of beam becomes an elliptical beam.

      (d) Too many lenses are used in the optical system.

      (e) Different wavelengths have different wave speeds.

10.   A biconvex lens is formed by using a piece of glass ($n = 1.52$). The radius of the front surface is 30 cm
      and the radius of the back surface is 40 cm. What is the focal length of the lens?

      (a) 2.3 m,   (b) –33 cm   (c) 33 cm   (d) 3.0 m   (e) –3.0 m

11.   Which one of the following describes the image of a convex lens when the object distance from the lens is
      smaller than the focal length?

      (a) virtual, upright, and magnified   (b) real, inverted, and reduced

      (c) virtual, upright, and reduced   (d) real, inverted, and magnified

      (e) virtual, inverted, and magnified

12.   A 3.0-cm-tall object is placed 12 cm in front of a mirror. A 9.0-cm-tall upright image is formed. What is
      the focal length of the mirror?

      (a) 36 cm   (b) 18 cm   (c) 54 cm   (d) 90 cm   (e) 72 cm

Answers to Practice Quiz:

1. b 2. b 3. e 4. c 5. a 6. b 7. e 8. b 9. a 10. c 11. a 12. b

CHAPTER 24

Physical Optics: The Wave Nature of Light

I. Chapter Objectives

Upon completion of this chapter, you should be able to:

1. explain how Young's experiment demonstrated the wave nature of light, and compute the wavelength of light from experimental results.
2. describe how thin films can produce colorful displays, and give some examples of practical applications of thin-film interference.
3. define diffraction, and give examples of diffractive effects.
4. explain light polarization, and give examples of polarization, both in the environment and in commercial applications.
*5. discuss scattering, and explain why the sky is blue and sunsets are red.

II. Chapter Summary and Discussion

1. Young's Double-Slit Experiment (Section 24.1)

Physical (wave) optics treats light as a wave in the study of some light phenomena, such as interference, diffraction, and polarization. These effects cannot be successfully explained with geometric optics, in which light is treated as several rays that follow straight-line paths.

Young's double-slit experiment not only demonstrates the wave nature of light but also allows the measurement of the wavelength of light. In this experiment, a light source is incident first on a single slit, then on two closely spaced narrow slits. The light waves emanating from these two slits can be considered as two in-phase spherical sources, and they interfere when they arrive at a screen. On the screen, the interference pattern consists of equally spaced maxima separated by equally spaced minima.

As for sound waves (Chapter 14), the condition for interference is determined by the *path-length difference* (ΔL) of the two waves, or the difference in distance traveled. If $\Delta L = n\lambda$ for $n = 0, 1, 2, 3, \ldots$, the interference is constructive (maximum); if $\Delta L = \dfrac{m\lambda}{2}$ for $m = 1, 3, 5, \ldots$, the interference is destructive (minimum). For two slits separated by a distance d, the path difference is approximately $\Delta L = d \sin \theta$ if the distance from the slits to the screen, L, is much greater than the slit separation, d. (This is often the case.) Thus, the condition for constructive interference in Young's double-slit experiment is $d \sin \theta = n\lambda$, for $n = 0, 1, 2, 3, \ldots$, where n is called the order number. The zeroth-order maximum ($n = 0$) corresponds to the central maximum, the first-order maximum ($n = 1$) is the first maximum on either side of the central maximum (there are two first-order maxima), and so on.

For a small angle θ, we have $\sin \theta \approx \tan \theta \approx \frac{y}{L}$, where y is the distance from the central maximum on the screen and L is the distance from the slits to the screen. The distance of the nth maximum (y_n) from the central maximum on either side is $y_n \approx \frac{nL\lambda}{d}$, the wavelength of light is then $\lambda \approx \frac{y_n d}{nL}$ for $n = 1, 2, 3, \ldots$, and the separation between adjacent maxima is $y_{n+1} - y_n = \frac{L\lambda}{d}$. (The minima are separated by this distance also.)

Example 24.1   Light of wavelength 632.8 nm falls on a double-slit, and the third-order maximum is seen at an angle of 6.5°. Find the separation between the double slits.

Solution: Given: $\lambda = 632.8 \text{ nm} = 632.8 \times 10^{-9} \text{ m}$, $n = 3$, $\theta_3 = 6.5°$.   Find: d.

From the condition for constructive interference, $d \sin \theta = n\lambda$, we have

$$d = \frac{n\lambda}{\sin \theta} = \frac{(3)\lambda}{\sin \theta_3} = \frac{(3)(632.8 \times 10^{-9} \text{ m})}{\sin 6.5°} = 1.7 \times 10^{-5} \text{ m} = 17 \ \mu\text{m}.$$

Example 24.2   In a Young's double-slit experiment, if the separation between the two slits is 0.10 mm, and the distance from the slits to a screen is 2.5 m, find the spacing between the first-order and second-order maxima for light with wavelength of 550 nm.

Solution: Given: $d = 0.10 \text{ mm} = 0.10 \times 10^{-3} \text{ m}$, $L = 2.5 \text{ m}$, $\lambda = 550 \text{ nm} = 550 \times 10^{-9} \text{ m}$.

Find: $y_2 - y_1$.

The position of the nth maximum measured form the central maximum is given by $y_n = \frac{nL\lambda}{d}$.

So $y_2 - y_1 = \frac{(2)L\lambda}{d} - \frac{(1)L\lambda}{d} = \frac{L\lambda}{d} = \frac{(2.5 \text{ m})(550 \times 10^{-9} \text{ m})}{0.10 \times 10^{-3} \text{ m}} = 1.4 \times 10^{-2} \text{ m} = 1.4 \text{ cm}.$

2. Thin-Film Interference (Section 24.2)

The waves reflected from the two surfaces of a thin film can interfere, and this **thin-film interference** depends on the reflective phases of the two reflected waves. Light reflected off a material whose index of refraction is greater than that of the one it is in $(n_2 > n_1)$ undergoes a *180° phase change*. If $n_2 < n_1$, there is *no phase change* on reflection. For a particular film, light reflected from the film surfaces may interfere constructively and destructively, depending on the film thickness and any additional phase changes. In most cases, the light is under normal (perpendicular) incidence, so the path length difference between the two waves reflected off the top and bottom surfaces of the film is $\Delta L = 2t$, where t is the film thickness; however, it is necessary to include any phase change in reflection when determining the conditions for constructive and destructive interferences.

A practical application of thin-film interference is the nonreflecting coatings for lenses (destructive interference for reflection of certain wavelengths). In a nonreflecting coating, the index of refraction of the film is usually between those of air and of glass ($n_o < n_1 < n_2$, where the subscript o is for air, 1 is for film, and 2 is for glass.), so there are 180° phase changes for each reflection off the two surfaces of the film. The minimum thickness of the film is given by $t_{min} = \dfrac{\lambda}{4n_1}$, where n_1 is the index of refraction of the film.

Other thin-film interference examples are **optical flats** and **Newton's rings**. An optical flat is a very flat and smooth piece of glass (usually the surface roughness is only $\lambda/20$). When a perfect spherical lens is placed on an optical flat, circular maxima and minima are observed. These are called **Newton's rings**. If a lens has irregularities, the rings are distorted. This is a simple, yet effective, method for checking the quality of lenses in optical industries. At the center of the Newton's rings where $t = 0$ (zero film thickness), a dark spot is observed. This is the direct evidence of the 180° phase shift, as normally a zero path length difference would result in constructive interference. Because it is a destructive interference, the two waves must be different by 180° in phase.

Example 24.3   A transparent material with an index of refraction 1.30 is used to coat a piece of glass with an index of refraction 1.52. What is the minimum thickness of the film in order to minimize the reflected light with a wavelength of 550 nm if the light is incident perpendicularly?

Solution: Given: $\lambda = 550$ nm, n_1 (film) = 1.30.   Find: t_{min} (minimum thickness).

The index of refraction of the film is between those of air and of glass ($n_o < n_1 < n_2$), so both reflections from the two surfaces of the film have 180° phase shifts. We can use Eq. (24.7).

The minimum film thickness is $t_{min} = \dfrac{\lambda}{4n_1} = \dfrac{550 \text{ nm}}{4(1.30)} = 106$ nm.

---

3. Diffraction (Section 24.3)

Diffraction is the deviation or bending of light around objects, edges, or corners. Generally, the smaller the size of the opening or object compared with the wavelength of light, the greater the diffraction. The diffraction pattern from a single slit of width w consists of a broader central maximum and some narrower side maxima (the width of the central maximum is twice that of the side maxima). Between two maxima, there is a region of destructive interference (minimum). In diffraction, the *minima rather than the maxima* are analyzed.

The condition for the minima is given by $w \sin \theta = m\lambda$, for $m = 1, 2, 3, \ldots$, where θ is the angle for a particular minimum designated by $m = 1, 2, 3, \ldots$ on either side of the central maximum (there is no minimum corresponding to $m = 0$. Why?) For a small-angle approximation, the position of the minimum from the center of the central maximum on a screen can be calculated from $y_m = \dfrac{mL\lambda}{w}$, where L is the distance from the single slit to the screen. The width of the mth side maximum is the distance between the mth minimum and the $(m+1)$th minimum, or $y_{m+1} - y_m$. The width of the central maximum is then $(y_1 - y_{-1})$ or $2(y_1 - y_o) = 2y_1$ (because $y_o = 0$). It is evident from the preceding discussions that

- for a given slit width (w), the greater the wavelength (λ), the wider the diffraction pattern (y_m);
- for a given wavelength (λ), the narrower the slit width (w), the wider the diffraction pattern (y_m);
- the width of the central maximum is twice the width of the side maxima.

Example 24.4   Light of wavelength 632.8 nm is incident on a single slit of width 0.200 mm. An observing screen is placed 2.50 m from the single slit. Find (a) the width of the central maximum and (b) the angle and positions of the third-order minimum.

Solution: Given: $\lambda = 632.8$ nm $= 632.8 \times 10^{-9}$ m, $w = 0.200$ mm $= 0.200 \times 10^{-3}$ m, $L = 2.50$ m,
                 Find: (a) $2y_1$ (b) θ_3 and y_3.

(a) The central maximum is the region between the first-order minima on either side of this maximum, so its width is simply $2y_1$.

From $y_m = \dfrac{mL\lambda}{w}$, we have $y_1 = \dfrac{(1)L\lambda}{w} = \dfrac{(1)(2.50 \text{ m})(632.8 \times 10^{-9} \text{ m})}{0.200 \times 10^{-3} \text{ m}} = 7.91 \times 10^{-3}$ m $= 7.91$ mm.

Thus, the width of the central maximum is $2y_1 = 2(7.91 \text{ mm}) = 15.8$ mm.

(b) Because $w \sin \theta = m\lambda$, $\sin \theta_3 = \dfrac{m\lambda}{w} = \dfrac{(3)(632.8 \times 10^{-9} \text{ m})}{0.200 \times 10^{-3} \text{ m}} = 9.49 \times 10^{-3}$.

So $\theta_3 = \sin^{-1}(9.49 \times 10^{-4}) = 0.544°$. $y_3 = \dfrac{(3)L\lambda}{w} = \dfrac{(3)(2.50 \text{ m})(632.8 \times 10^{-9} \text{ m})}{0.200 \times 10^{-3} \text{ m}} = 23.7$ mm.

Example 24.5 Light of wavelength 550 nm is incident on a 0.75-mm-wide single slit. At what distance from the single slit should a screen be placed if the second-order minimum in the diffraction pattern is to be 1.7 mm from the center of the screen?

Solution: Given: $\lambda = 550$ nm $= 550 \times 10^{-9}$ m, $w = 0.75$ mm $= 0.75 \times 10^{-3}$ m, $m = 2$,
$y_2 = 1.7$ mm $= 1.7 \times 10^{-3}$ m.

Find: L.

From $y_m = \dfrac{mL\lambda}{w}$, we have $L = \dfrac{y_m w}{m\lambda} = \dfrac{(1.7 \times 10^{-3}\ \text{m})(0.75 \times 10^{-3}\ \text{m})}{(2)(550 \times 10^{-9}\ \text{m})} = 1.2$ m.

A **diffraction grating** consists of a larger number of closely spaced narrow slits. The diffraction pattern of a diffraction grating is a combination of multiple-slit interference and single-slit diffraction. A diffraction grating is very useful for dispersing different wavelengths or colors. (In general, it gives a much larger dispersion than a prism.) The angular positions of the sharp (narrow, or well-defined) maxima for a diffraction grating are given by $d \sin \theta = n\lambda$, for $n = 0, 1, 2, 3, \ldots$, where d is the spacing between adjacent grating slits (or grating spacing), which can be obtained from the number of lines per unit length of the grating, $d = 1/N$. The number of spectral orders produced by a grating depends on the wavelength and on the grating's spacing d. Because $\sin \theta$ cannot be greater than 1, $\sin \theta = (n\lambda)/d \le 1$, so the order number is therefore limited by $n \le d/\lambda$.

Note: In general, the small-angle approximation cannot be used in a diffraction grating because of the grating's larger dispersion power. The maxima are usually separated by large angles, and, therefore, the angles do not satisfy the small-angle approximation.

Integrated Example 24.6

Monochromatic light is incident on two diffraction gratings that have 4000 lines/cm and 5000 lines/cm, respectively. The third-order maximum of the 5000 lines/cm grating is seen at 46.3°. (a) The third-order maximum of the 4000 lines/cm grating should be seen at an angle that is (1) greater than, (2) equal to, or (3) less than 46.3°. Explain. (b) Calculate the angle for the third-order maximum of the 4000 lines/cm grating and the wavelength of the incident light?

(a) Conceptual Reasoning:

Since the 4000 lines/cm grating has fewer lines per centimeter, it has a greater grating constant d (the distance between adjacent lines). According to the maximum for a diffraction grating, $d \sin \theta = n\lambda$, a greater d corresponds to a smaller $\sin \theta$ and therefore θ, for a given order number n and wavelength λ. Thus, the third-order maximum of the 4000 lines/cm grating should be seen at an angle that is less than 46.3°.

(b) Quantitative Reasoning and Solution:

We identify the 4000 lines/cm grating as grating A, the 5000 lines/cm grating as grating B, and use subscripts A and B to denote the two gratings.

Given: $n = 3$, $(\theta_3)_B = 46.3°$, $N_A = 4000$ lines/cm, $N_B = 5000$ lines/cm.

Find: $(\theta_3)_A$ and λ.

The grating constants are

$$d_A = \frac{1}{N_A} = \frac{1}{4000 \text{ lines/cm}} = 2.50 \times 10^{-4} \text{ cm} = 2.50 \times 10^{-6} \text{ m}.$$

$$d_B = \frac{1}{N_B} = \frac{1}{5000 \text{ lines/cm}} = 2.00 \times 10^{-4} \text{ cm} = 2.00 \times 10^{-6} \text{ m}.$$

From $d \sin \theta = n\lambda$, we have $d_A \sin (\theta_3)_A = n\lambda$, and $d_B \sin (\theta_3)_B = n\lambda$.

Since n and λ are the same for both gratings, $\dfrac{d_A \sin (\theta_3)_A}{d_B \sin (\theta_3)_B} = \dfrac{n\lambda}{n\lambda} = 1$.

Therefore $\sin (\theta_3)_A = \dfrac{d_B \sin (\theta_3)_B}{d_A} = \dfrac{(2.00 \times 10^{-6} \text{ m}) \sin 46.3°}{2.50 \times 10^{-6} \text{ m}} = 0.578.$

Thus $(\theta_3)_A = \sin^{-1}(0.578) = 35.3°$. As expected, the angle for the third-order maximum is smaller for the 4000 lines/cm grating.

We can use either grating to calculate the wavelength. Here we use the 5000 lines/cm grating.

$$\lambda = \frac{d \sin \theta}{n} = \frac{(2.00 \times 10^{-6} \text{ m}) \sin 46.3°}{3} = 4.82 \times 10^{-7} \text{ m} = 482 \text{ nm}.$$

Example 24.7   Monochromatic light of wavelength 632.8 nm is incident normally on a diffraction grating. If the second-order maximum of the diffraction pattern is observed at 32.0°,

(a) what is the grating spacing?

(b) how many total number of visible maxima can be seen?

Solution: Given: $\lambda = 632.8 \text{ nm} = 632.8 \times 10^{-9} \text{ m}$, $n = 2$, $\theta_2 = 32.0°$.

Find: (a) d (b) total number of visible maxima.

(a) From $d \sin \theta = n\lambda$, we have $d = \dfrac{n\lambda}{\sin \theta} = \dfrac{(2)(632.8 \times 10^{-9} \text{ m})}{\sin 32.0°} = 2.39 \times 10^{-6} \text{ m}.$

(b) The maximum value of the angle θ is $90°$ ($\sin 90° = 1$), so the maximum order number is

$$n_{max} = \frac{d}{\lambda} = \frac{2.39 \times 10^{-6} \text{ m}}{632.8 \times 10^{-9} \text{ m}} = 3.78. \quad \text{Because } n \text{ is an integer, } n_{max} = 3.$$

Therefore, 7 maxima are seen including the center one. (Why?)

(The 7 maxima include one for $n = 0$, two for $n = 1$, two for $n = 2$, and two for $n = 3$.)

The regular atomic spacing in a crystalline solid acts as a diffraction grating for light of much shorter wavelength than visible light, such as X-rays. From a measurement of the diffraction angle, which is equal to the incidence angle also, of an X-ray beam with known wavelength, the distance between the crystal's internal planes (d) can be determined from **Bragg's law**, $2d \sin \theta = n\lambda$, for $n = 1, 2, 3, \ldots$.

4.   Polarization (Section 24.4)

Polarization is the preferential orientation of the electromagnetic field vectors that make up a light wave and is evidence that light is a transverse wave. Light with some partial preferential orientation of the electromagnetic field vectors is said to be *partially polarized*. If the electromagnetic field vectors oscillate in *one* plane (or *one* direction), the light is then *plane (linearly) polarized*. Light can be polarized by selective absorption (**dichroism**), reflection, double refraction (**birefringence**), and scattering.

Some crystals, such as tourmaline and herapathite, exhibit the interesting property of absorbing one of the polarized components more than the other. This property is called dichroism. If the dichroic material is sufficiently thick, the more strongly absorbed component may be completely absorbed, resulting in a linearly polarized beam with the unabsorbed component. Polaroid films (used in sunglasses) are synthetic polymer materials that allow light of one polarization direction to pass. This direction is called the **transmission axis**, or **polarization direction**. If two films (the first one is called the polarizer, and the second one is called the analyzer) are placed with their transmission axes parallel to each other, light can pass through both films. If the two transmission axes are perpendicular to each other, little or no light can pass. In general, the intensity of the transmitted light is given by $I = I_0 \cos^2 \theta$, where θ is the angle between the transmission axes of the polarizer and analyzer, I_0 is the intensity after the (first) polarizer, and I is the intensity after the analyzer. This expression is known as *Malus's law*.

Example 24.8   Unpolarized light is incident on a polarizer-analyzer pair. The transmission axes of the polarizer and analyzer are at 35° to each other. What percentage of the light gets through the pair?

Solution:

When unpolarized light is incident on a polarizer, only one of the two electric field components can pass; that is, only 50% of the intensity gets through the (first) polarizer. Thus, before the light reaches the (second) analyzer, the intensity is already reduced by half.

From Malus's law, $I = I_0 \cos^2 \theta$, $\dfrac{I}{I_0} = \cos^2 \theta = \cos^2 35° = 0.671$ or 67.1%.

Therefore, only $\dfrac{67.1\%}{2} = 34\%$ of the original intensity passes through the polarizer-analyzer pair.

Light that is partially reflected and partially refracted is partially polarized; however, when the reflected and the refracted rays form a 90° angle, the reflected ray is linearly polarized, or maximum polarization occurs. The angle of incidence for this maximum polarization is called the **polarizing** or **Brewster angle**. The polarizing angle for reflection occurs at material interface of indices of refraction n_1 and n_2 and is given by $\tan \theta_p = n_2/n_1$. If the first material is air ($n_1 = 1$), then $\theta_p = \tan^{-1} n_2 = \tan^{-1} n$, where n is the index of refraction of the second material.

Example 24.9   How far above the horizon is the Moon when its image reflected in calm water is completely polarized?

Solution: Given: $n_1 = 1$ (air), $n_2 = 1.33$ (water).   Find: $90° - \theta_p$.

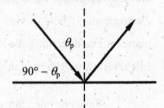

The polarizing angle is the angle of incidence (measured from the normal) when the reflected ray is linearly polarized. The angle above the horizon is the angle measured from the surface ($90° - \theta_p$).

From $\tan \theta_p = n_2/n_1$, we have $\theta_p = \tan^{-1} n_2/n_1 = \tan^{-1} 1.33/1 = 53.1°$.

So the angle above the horizon is $90° - 53.1° = 36.9°$.

In some materials, the anisotropy of the speed of light with direction (speed of light is different in different directions) gives rise to different indices of refraction in different directions. This property is called *birefringence*, and such materials are said to be *birefringent* or *double refracting*. When a beam of unpolarized light is incident on a birefringent crystal, it is doubly refracted and separated into two components, or rays. These two rays are linearly polarized, with the electromagnetic field vectors in mutually perpendicular directions.

Some transparent materials have the ability to rotate the polarization direction of the linearly polarized light. This phenomenon is called **optical activity** and is due to the molecular structure of the material. Some liquid crystals are optically active, and this property forms the basis of the common **liquid crystal display (LCD)**.

*5.   Atmospheric Scattering of Light (Section 24.5)

Scattering is the process in which particles (like air molecules or dust particles) absorb light and reradiate polarized light. The scattering of sunlight by air molecules causes the sky to look blue and sunsets to look red because the shorter-wavelength (blue) light is scattered more efficiently than longer-wavelength (red) light. This is called **Rayleigh scattering**, and the scattering intensity is found to be proportional to $1/\lambda^4$.

Blue scatters more efficiently than red. In the morning and evening, the blue component of the light from the Sun is scattered more in the denser atmosphere near the Earth, so we see red when we look in the direction of the rising or setting Sun. During the day, we mainly see the blue component from overhead scattering.

III. Mathematical Summary

Angles for Maxima (Double-Slit)	$d \sin \theta = n\lambda$ for $n = 0, 1, 2, \ldots$ (24.3)	Gives the angular positions of the maxima in a double-slit interference experiment.
Lateral Position for Maxima (Double-Slit)	$y_n \approx \dfrac{nL\lambda}{d}$ for $n = 0, 1, 2, \ldots$ (24.4)	Gives the approximate lateral positions of the maxima in a double-slit interference experiment.
Minimum Thickness for a Nonreflecting Film	$t_{min} = \dfrac{\lambda}{4n_1}$ (for $n_2 > n_1 > n_0$) (24.7)	Gives the minimum thickness of a thin-film coating to minimize reflection.
Angles for Minima (Single-Slit)	$w \sin \theta = m\lambda$ for $m = 1, 2, 3, \ldots$ (24.8)	Gives the angular positions of the minima in a single-slit diffraction experiment.
Angles for Maxima (Diffraction Grating)	$d \sin \theta = n\lambda$ for $n = 0, 1, 2, \ldots$ (24.12)	Determines the angular positions of the maxima for a diffraction grating. ($d = 1/N$ and N is the number of lines per unit length.)
Malus's Law	$I = I_0 \cos^2 \theta$ (24.14)	Calculates the intensity of light after a polarizer-analyzer pair whose transmission axes are at an angle θ.
Brewster (polarizing) Angle	$\tan \theta_p = n_2/n_1$ or $\theta_p = \tan^{-1} n_2/n_1$ (24.15)	Calculates the polarizing or Brewster angle in a boundary between two media.

IV. Solutions of Selected Exercises and Paired/Trio Exercises

8. $0.75 \text{ m} = 0.50 \text{ m} + 0.25 \text{ m} = 1.5(0.50 \text{ m}) = 1.5\lambda$. So the waves will interfere $\boxed{\text{destructively}}$.

    $1.0 \text{ m} = 0.50 \text{ m} + 0.50 \text{ m} = 2(0.50 \text{ m}) = 2\lambda$. So the waves will interfere $\boxed{\text{constructively}}$.

12. (a) Since $\Delta y = \dfrac{L\lambda}{d}$, Δy is proportional to λ ($\Delta y \propto \lambda$). Therefore the distance between the maxima will

    $\boxed{(1) \text{ increase}}$ if λ increases.

    (b) $y_n \approx \dfrac{nL\lambda}{d}$, the distance is equal to $\Delta y = y_3 - y_0 = \dfrac{3L\lambda}{d} = \dfrac{3(1.5 \text{ m})(550 \times 10^{-9} \text{ m})}{0.25 \times 10^{-3} \text{ m}} = \boxed{0.99 \text{ cm}}$.

    (c) $\Delta y = y_3 - y_0 = \dfrac{3L\lambda}{d} = \dfrac{3(1.5 \text{ m})(680 \times 10^{-9} \text{ m})}{0.25 \times 10^{-3} \text{ m}} = \boxed{1.2 \text{ cm}} > 0.99 \text{ cm}$.

16.　(a) The answer is $\boxed{(1)\ \text{increase}}$. Since $\Delta y = \dfrac{L\lambda}{d}$, Δy is proportional to L ($\Delta y \propto L$), the distance between the maxima will also increase if the distance from the double slits to the screen is increased.

(b) $\Delta y = \dfrac{L\lambda}{d} = \dfrac{(2.00\ \text{m})(550 \times 10^{-9}\ \text{m})}{1.75 \times 10^{-4}\ \text{m}} = \boxed{0.63\ \text{cm}}$.

(c) $\Delta y = \dfrac{(3.00\ \text{m})(550 \times 10^{-9}\ \text{m})}{1.75 \times 10^{-4}\ \text{m}} = \boxed{0.94\ \text{cm}} > 0.63\ \text{cm}.$

18.　(a) The answer is $\boxed{(3)\ \text{decreases}}$. If it is in water, $\lambda' = \dfrac{\lambda}{n} = \dfrac{\lambda}{1.33}$ or the wavelength decreases.

Also $\Delta y = \dfrac{L\lambda}{d}$ so Δy is proportional to λ ($\Delta y \propto \lambda$), the spacing of the interference fringes will decrease if wavelength decreases.

(b) For 24-12(b), $\Delta y' = \dfrac{\Delta y}{1.33} = \dfrac{0.99\ \text{cm}}{1.33} = \boxed{7.4\ \text{mm}}$

For 24-12(c), $\Delta y' = \dfrac{1.2\ \text{cm}}{1.33} = \boxed{9.0\ \text{mm}}$.

26.　(a) The answer is $\boxed{(3)\ \text{two}}$. Both waves experience $180°$ phase shift.

(b) The minimum thickness for an antireflection coating is $t_{min} = \dfrac{\lambda}{4n_1}$.

So $\lambda = 4n_1 t_{min} = 4(1.4)(1.0 \times 10^{-7}\ \text{m}) = 5.6 \times 10^{-7}\ \text{m} = \boxed{560\ \text{nm}}$.

28.　The minimum thickness for an antireflection coating is
$t_{min} = \dfrac{\lambda}{4n_1} = \dfrac{700\ \text{nm}}{4(1.40)} = \boxed{1.25\ \text{nm}}$.

34.　(a) The answer is $\boxed{(3)\ \text{bright and dark lines}}$. The two rays for interference are the reflections from the bottom surface of the top plate and the top surface from the bottom plate. As the distance between the top and bottom plates increases, so does the path length difference. The path-length difference will alternate between $n\lambda$ and $(n + \frac{1}{2})\lambda$, where n is an integer. Therefore, the interference pattern consists of equally spaced maxima and minima.

(a) The reflection from the top surface of the bottom plate has $180°$ phase shifts (half-wave), so the condition for constructive interference for reflection is $\Delta L = 2t + \dfrac{\lambda}{2} = m\lambda$, $m = 1, 2, 3, \dots$.

So $\boxed{2t = (m - \frac{1}{2})\lambda,\ m = 1, 2, 3, \dots}$.

42.   (a) The width of the central maximum is

$$2\Delta y = 2y_1 = \frac{2L\lambda}{w} = \frac{2(1.0 \text{ m})(480 \times 10^{-9} \text{ m})}{0.20 \times 10^{-3} \text{ m}} = \boxed{4.8 \text{ mm}}.$$

(b) The width of the side maxima is half the width of the central maximum.

$$\Delta y_3 = \Delta y_4 = \frac{L\lambda}{w} = \boxed{2.4 \text{ mm}}.$$

46.   (a) The answer is $\boxed{\text{(3) decrease}}$ because the width of the central maximum is equal to $2\Delta y = 2y_1 = \dfrac{2L\lambda}{w}$.

If w increases, the width decreases.

(b) $2\Delta y = 2y_1 = \dfrac{2L\lambda}{w} = \dfrac{2(1.80 \text{ m})(680 \times 10^{-9} \text{ m})}{0.50 \times 10^{-3} \text{ m}} = 4.9 \times 10^{-3} \text{ m} = \boxed{4.9 \text{ mm}}.$

(c) $2\Delta y = \dfrac{2(1.80 \text{ m})(680 \times 10^{-9} \text{ m})}{0.60 \times 10^{-3} \text{ m}} = 4.1 \times 10^{-3} \text{ m} = \boxed{4.1 \text{ mm}} < 4.9 \text{ mm}.$

50.   (a) The answer is $\boxed{\text{(3) both}}$. The maximum angle is $90°$.

From $d \sin \theta = n\lambda$, $n_{max} = \dfrac{d \sin \theta}{\lambda} = \dfrac{d \sin 90°}{\lambda} = \dfrac{d}{\lambda}$. So it depends on both d and λ.

(b) $d = \dfrac{1}{10\,000 \text{ /cm}} = 1.00 \times 10^{-4} \text{ cm} = 1.00 \times 10^{-6} \text{ m}$. $n_{max} = \dfrac{d}{\lambda} = \dfrac{1.00 \times 10^{-6} \text{ m}}{560 \times 10^{-9} \text{ m}} = 1.8.$

So n_{max} can only be 1 because n is an integer.

Therefore, the number of maxima that appear is $\boxed{3, n = 0, 1 \text{ (two of them)}}$.

52. $d = \dfrac{1}{4000 \text{ lines/cm}} = 2.5 \times 10^{-4} \text{ cm} = 2.5 \times 10^{-6} \text{ m}.$

Using $d \sin \theta = n\lambda$, we have $\theta = \sin^{-1} \dfrac{n\lambda}{d}$.

If they do overlap, it will be the first-order red to the second-order blue.

For blue, $\theta_{2b} = \sin^{-1} \dfrac{(2)(400 \times 10^{-9} \text{ m})}{2.5 \times 10^{-6} \text{ m}} = 18.7°.$

For red, $\theta_{1r} = \sin^{-1} \dfrac{700 \times 10^{-9} \text{ m}}{2.5 \times 10^{-6} \text{ m}} = 16.3°.$

So $\theta_{2b} > \theta_{1r}$; so the answer is $\boxed{\text{no}}$, they do not overlap.

56.   (a) The answer is $\boxed{\text{(2) only the boy at } 19.6°}$ may not hear the whistle. The boy at $0°$ is at the position of

the central maximum ($\theta = 0°$) so he will always hear the whistle. The boy at a9.6°, however, may be in a

diffraction minimum so he may not hear the whistle.

(b) $\lambda = \dfrac{v}{f} = \dfrac{335 \text{ m/s}}{1000 \text{ Hz}} = 0.335$ m. Using $w \sin \theta = m\lambda$, we have $\theta = \sin^{-1} \dfrac{m\lambda}{w}$.

$\theta_1 = \sin^{-1} \dfrac{0.335 \text{ m}}{1.0 \text{ m}} = 19.6°$. So there is a minimum at 19.6°, and therefore the answer is $\boxed{\text{no}}$, the boy at 19.6° cannot hear the whistle.

61. When the axes are perpendicular, it darkens, and when the axes are parallel it lightens.

   (a) $\boxed{\text{Twice}}$.   (b) $\boxed{\text{Four times}}$.

   (c) $\boxed{\text{None}}$.   (d) $\boxed{\text{Six times}}$.

66. $I = I_0 \cos^2 \theta.$ For $\theta = 30°$, $I = I_0 \cos^2 30° = 0.75 I_0.$

                                         For $\theta = 45°$, $I = I_0 \cos^2 45° = 0.50 I_0.$

   So $\boxed{30°}$ will allow more light to be transmitted.

70. The answer is $\boxed{\text{(3) some light}}$. The fact that the polarization is maximum in the reflected light does not mean that the light intensity is maximum in the reflected light. There will still be transmitted light, which is also polarized, but the polarization direction is perpendicular to that of the reflected light.

   (b) Since $n_1 = 1$ (air), $\tan \theta_p = n_2/n_1 = n_2$. So $\theta_p = \tan^{-1} n_2$.

   Therefore $\theta_1 = \theta_p = \tan^{-1} n_2 = \tan^{-1} 1.22 = 50.7°$.

   Also from Snell's law, $n_1 \sin \theta_1 = n_2 \sin \theta_2$, we have $\sin \theta_2 = \dfrac{n_1 \sin \theta_1}{n_2} = \dfrac{(1) \sin 50.7°}{1.22} = 0.634.$

   So $\theta_2 = \sin^{-1} (0.634) = \boxed{39.3°}$.

72. (a) The answer is $\boxed{\text{(2) less than}}$. Since $\theta_p = \tan^{-1} n_2/n_1$ with $n_1 = 1$ (air) and $n_2 = 1.60$ or $n_1 = 1.33$ (water) and $n_2 = 1.60$, n_2/n_1 is smaller for water. So θ_p in water is less than θ_p in air.

   (b) In air, $\theta_p = \tan^{-1} n_2/n_1 = \tan^{-1} 1.60/1.00 = \boxed{58.0°}$.

   In water, $\theta_p = \tan^{-1} n_2/n_1 = \tan^{-1} 1.60/1.33 = \boxed{50.3°} < 58.0°.$

76. $\boxed{\text{Blue scatters more efficiently than red.}}$. In the morning and evening, the blue component of the light from the Sun is scattered more in the denser atmosphere near the Earth, so we see red when we look in the direction of the rising or setting Sun. During the day, we mainly see the scattered blue component from overhead.

79. Since $n_2 = 1$ (air), $\tan \theta_p = n_2/n_1 = 1/n_1$. For total internal reflection, $\sin \theta_c = n_2/n_1 = 1/n_1$.

That means $\tan \theta = \sin \theta$. This is not possible for any angle that is not equal to zero.

So the answer is $\boxed{no}$.

(b) $1/n_1 = \sin \theta_c = \sin 35° = 0.5736$. $\tan \theta_p = 1/n_1 = 0.5736$. So $\theta_p = \boxed{29.8°}$.

83. $d = \dfrac{1}{9000 \text{ lines/cm}} = 1.11 \times 10^{-4} \text{ cm} = 1.11 \times 10^{-6} \text{ m}$.

From $d \sin \theta = n\lambda$, we have $n_{max} = \dfrac{d \sin 90°}{\lambda} = \dfrac{d}{\lambda}$.

For red, $n_{max} = \dfrac{1.11 \times 10^{-6} \text{ m}}{700 \times 10^{-9} \text{ m}} = 1.6$. So $n_{max} = \boxed{1 \text{ for red}}$.

For violet, $n_{max} = \dfrac{1.11 \times 10^{-6} \text{ m}}{400 \times 10^{-9} \text{ m}} = 2.8$. So $n_{max} = \boxed{2 \text{ for violet}}$.

V. Practice Quiz

1. If a wave from one slit of a Young's double-slit experiment arrives at a point on the screen two wavelengths behind the wave from the other slit, what is observed at that point?

(a) bright fringe   (b) dark fringe   (c) gray fringe   (d) multi-colored fringe   (e) none of the above

2. Monochromatic light is incident on a Young's double-slit separated by 3.00×10^{-5} m. The resultant bright fringe separation is 2.15×10^{-2} m on a screen 1.20 m from the double slit. What is the separation between the third-order bright fringe and the zeroth-order fringe?

(a) 8.60×10^{-2} m   (b) 7.35×10^{-2} m   (c) 6.45×10^{-2} m   (d) 4.30×10^{-2} m   (e) 2.15×10^{-2} m

3. What is the minimum thickness of a nonreflecting coating ($n = 1.35$) on a glass lens ($n = 1.52$) for wavelength 550 nm?

(a) zero   (b) 102 nm   (c) 204 nm   (d) 90.5 nm   (e) 181 nm

4. What will happen to the width of the central maximum if the width of the single slit decreases in a single-slit experiment?

(a) It will decrease.   (b) It will increase.   (c) It will remain unchanged.

(d) It does not depend on the separation.

(e) The answer cannot be determined because not enough information is given.

5. Light of wavelength 610 nm is incident on a slit 0.20-mm wide, and the diffraction pattern is produced on a screen that is 1.5 m from the slit. What is the width of the central maximum?

    (a) 0.34 cm   (b) 0.68 cm   (c) 0.92 cm   (d) 1.22 cm   (e) 1.35 cm

6. A beam of unpolarized light in air strikes a flat piece of glass at an angle of incidence of 57.0°. If the reflected beam is completely polarized, what is the index of refraction of the glass?

    (a) 0.54   (b) 0.84   (c) 1.12   (d) 1.54   (e) 1.84

7. What is the process responsible for producing polarized light in a dichroic material like Polaroid film?

    (a) reflection   (b) refraction   (c) double refraction   (d) selective absorption   (e) scattering

8. When the transmission axes (polarization directions) of two polarizers are parallel to each other, what percentage of the incident light will pass the two sheets if the incident light is unpolarized?

    (a) 0%   (b) 25%   (c) 50%   (d) 75%   (e) 100%

9. White light is spread out into spectral hues by a diffraction grating. If the grating has 2000 lines per centimeter, at what angle will red light ($\lambda = 640$ nm) appear in the first order?

    (a) 0°   (b) 3.57°   (c) 7.35°   (d) 11.2°   (e) 13.4°

10. A helium–neon laser ($\lambda = 632.8$ nm) is used to calibrate a diffraction grating. If the first-order maximum occurs at 20.5°, how many lines are there in a millimeter?

    (a) 138   (b) 185   (c) 276   (d) 455   (e) 553

11. Monochromatic light of wavelength 550 nm falls on two slits separated by a distance of 40.0 μm. What is the distance between the first-order and the fifth-order bright fringes formed on a screen 1.00 m away from the slits?

    (a) 1.25 cm   (b) 2.50 cm   (c) 3.75 cm   (d) 5.00 cm   (e) 6.25 cm

12. A film on a lens is 1.0×10^{-7} m thick and is illuminated with white light. The index of refraction of the film is 1.35, and the index of refraction of the glass is 1.51. For what wavelength of light will the lens be nonreflecting?

    (a) 135 nm   (b) 270 nm   (c) 405 nm   (d) 540 nm   (e) 675 nm

Answers to Practice Quiz:

1. a 2. c 3. b 4. b 5. c 6. d 7. d 8. c 9. c 10. e 11. d 12. d

CHAPTER 25

Vision and Optical Instruments

I.   Chapter Objectives

Upon completion of this chapter, you should be able to:

1.   describe the optical workings of the eye, and explain some common vision defects and how they are corrected.

2.   distinguish between lateral and angular magnifications, and describe simple and compound microscopes and their magnifications.

3.   distinguish between refractive and reflective telescopes, and describe the advantages of each.

4.   describe the relationship of diffraction and resolution, and state and explain Rayleigh's criterion.

*5.   relate color vision and light.

II.   Chapter Summary and Discussion

1.   The Human Eye (Section 25.1)

The crystalline lens in the eye is a converging lens composed of microscopic glassy fibers. Through muscle action, the shape of the lens is adjusted (this adjustment is called *accommodation*), and sharp images are formed on the **retina**, a light-sensitive surface at the back of the eye. The photo-sensitive **rod** and **cone** cells of the retina are responsible for twilight (black-and-white) vision and color vision, respectively.

The extremes of the range over which distinct vision (sharp focus) is possible are known as the *far point* and the *near point*. The far point is the greatest distance at which the eye can see objects clearly, and is for the normal eye taken to be infinity. The near point is the position closest to the eye at which objects can be seen clearly and depends on the extent the lens can be deformed (thickened) by accommodation, which varies with age even for the normal eye.

There are three common vision defects. **Nearsightedness** is the condition of being able to see nearby objects clearly, but not distant objects, because the focal length of the crystalline lens is too short (the lens is too converging; the image is focused in front of the retina). A person with nearsightedness will have the far point not at infinity but at a nearer point. This vision defect can be corrected by using a diverging lens. **Farsightedness** is the condition of being able to see distant objects clearly, but not nearby objects, because the focal length of the

---

crystalline lens is too long (the lens is too diverging; the image is focused behind the retina). A person with farsightedness will have the near point not at the normal position but at some point farther from the eye. This vision defect can be corrected by using a converging lens. **Astigmatism** is caused by a refractive surface, most usually the cornea or crystalline lens, being out of spherical shape. As a result, the eye has different focal lengths in different planes. This condition may be corrected with lenses that have greater curvature in the plane in which the cornea or crystalline lens has deficient curvature.

Integrated Example 25.1

A student cannot see clearly objects more than 0.950 m away. (a) The corrective lens should be (1) converging, (2) flat, or (3) diverging. Explain. (b) What should be the powers of the lenses (regular glass and contact lens) prescribed if the glass is to be worn 1.00 cm in front of the eye?

(a) Conceptual Reasoning:

Since the student cannot see things farther than 0.950 m, the student has nearsightedness. To correct nearsightedness, the corrective lens should be (3) diverging (negative focal length or negative power) so it can bring the image of an object at infinity to the far point of the student.

(b) Quantitative Reasoning and Solution:

Given: $d_o = \infty$, $d_i = -(0.950\text{ m} - 0.010\text{ m}) = -0.940\text{ m}$ (regular glass),

$d_i = -0.950$ m (contact lens on the eye).

Find: $P = \dfrac{1}{f}$ for both glass and contact lens.

A normal eye has a far point at infinity. To correct for nearsightedness, the image of an object at infinity must be formed at 0.950 m in front of the student's eye.

For regular glass, the image distance is $-(0.950\text{ m} - 0.0100\text{ m}) = -0.940$ m. The image distance is negative because the image is on the same side as the object (virtual image). Using the thin-lens equation,

$\dfrac{1}{f} = \dfrac{1}{d_o} + \dfrac{1}{d_i}$, we have

$$P = \frac{1}{f} = \frac{1}{d_o} + \frac{1}{d_i} = \frac{1}{\infty} + \frac{1}{-0.940\text{ m}} = 0 - 1.06\text{ m}^{-1} = -1.06\text{ D.}$$

For contact lens, the image distance is simply $d_i = -0.950$ m because the lens is right on the eye. Again, the image distance is negative because the image is on the same side as the object (virtual image).

$$P = \frac{1}{f} = \frac{1}{d_o} + \frac{1}{d_i} = \frac{1}{\infty} + \frac{1}{-0.950\text{ m}} = 0 - 1.05\text{ m}^{-1} = -1.05\text{ D.}$$

As expected, the powers of the glass and contact lens is negative (diverging lens).

---

2. Microscopes (Section 25.2)

A **magnifying glass (simple microscope)** is a single converging lens that allows one to view an object clearly when it is brought closer than the near point. In such a position, an object subtends a greater angle and therefore appears larger, or magnified. The magnification of an object viewed through a magnifying glass is expressed in terms of **angular magnification** (m), which is defined as the ratio of the *angular* size of the object viewed through the magnifying glass (θ) to the angular size of the object viewed without the magnifying glass (θ_0), or $m = \theta/\theta_0$. If the image is at the near point (normally 25 cm from the eye), the magnification of the magnifying glass is $m = 1 + (25 \text{ cm})/f$, where f is the focal length of the converging lens. If the image is formed at infinity (eye in relaxed position), the magnification is then $m = (25 \text{ cm})/f$. Note that the shorter the focal length, the greater the magnification. A magnifying glass provides limited magnification because for very short focal lengths, the image becomes distorted.

A **compound microscope** consists of a pair of converging lenses, each of which contributes to the magnification. A compound microscope provides greater magnification than can be attained with a single lens. A converging lens having a relatively short focal length ($f_o < 1$ cm) is known as the **objective**. It forms a real, inverted, and magnified image of an object positioned *slightly beyond* its focal point. The other lens, called the **eyepiece** (or **ocular**), has a longer focal length (f_e is a few centimeters) and is positioned so that the image formed by the objective falls just *inside* its focal point. This lens forms a magnified virtual image that is viewed by the observer. The final image is inverted in relation to the original object. The total angular magnification of the combination is $M_{total} = M_o m_e = -\dfrac{(25 \text{ cm}) L}{f_o f_e}$, where L is the separation between the two lenses, and f_o, f_e, and L are in centimeters. When finding the magnification of a compound microscope, make sure to convert f_o, f_e, and L into the same units, usually centimeters, because the near point is usually so expressed.

Example 25.2   A person uses a converging lens of focal length 5.0 cm as a magnifying glass.

(a) What is the maximum possible angular magnification?

(b) What is the magnification if the person's eye is relaxed?

Solution: Given: $f = 5.0$ cm.   Find: (a) m (maximum)   (b) m (relaxed).

(a) The maximum magnification is attained when the image is at the near point.

$$m = 1 + \frac{25 \text{ cm}}{f} = 1 + \frac{25 \text{ cm}}{5.0 \text{ cm}} = 6.0\times.$$

(b) When the eyes are relaxed, the image is at infinity, so $m = \dfrac{25 \text{ cm}}{f} = 5.0\times.$

Example 25.3 A compound microscope has an objective with a focal length of 4.5 mm and an eyepiece of focal length 5.0 cm. If the two lenses are separated by 20 cm, what is the total angular magnification?

Solution: Given: $f_o = 4.5$ mm = 0.45 cm, $f_e = 5.0$ cm, $L = 20$ cm.

Find: M_{total}.

$$M_{total} = \frac{(25\text{ cm})L}{f_o f_e} = \frac{(25\text{ cm})(20\text{ cm})}{(0.45\text{ cm})(5.0\text{ cm})} = -220\times.$$

The negative sign indicates that the final image is inverted in relation to the original object.

3. Telescopes (Section 25.3)

A **refracting telescope** uses a converging lens to collect and converge light from a distant object whereas a **reflecting telescope** uses a mirror to collect and converge light. The principle of a refracting telescope is similar to that of a compound microscope. The major components are objective and eyepiece lenses. The objective is a large converging lens with a long focal length. The movable eyepiece has a relatively short focal length. The image formed by the objective acts like the object for the eyepiece, and a magnified image is seen. If the eyepiece of a refracting telescope is also a converging lens, the final image is inverted, and the setup is known as an **astronomical telescope**. If the final image is upright, the setup is known as a **terrestrial telescope**. There are several ways to achieve an upright image. One of these is to use a diverging lens as the eyepiece (Galilean telescope). The separation of the objective and the eyepiece is equal to the sum of the focal lengths of the two lenses, $L = f_o + f_e$. The angular magnification of a refracting telescope for a final image at infinity is given by $m = -f_o/f_e$.

Example 25.4 A student constructs an astronomical telescope with a magnification of 10. If the telescope has a converging lens of focal length 50 cm,

(a) what should be the focal length of the eyepiece?

(b) what is the resulting length of the telescope?

Solution: An astronomical telescope has an inverted final image, so the magnification is negative.

Given: $m = -10$ (astronomical telescope forms inverted images), $f_o = 50$ cm.

Find: (a) f_e (b) L.

(a) From $m = -\dfrac{f_o}{f_e}$, we have $f_e = -\dfrac{f_o}{m} = -\dfrac{50\text{ cm}}{-10} = 5.0$ cm.

(b) The length of a refracting telescope is equal to the sum of the focal lengths of the two lenses.

$L = f_o + f_e = 50$ cm + 5.0 cm = 55 cm.

4.   Diffraction and Resolution (Section 25.4)

The diffraction of light places a limitation on our ability to distinguish objects that are close together when using microscopes or telescopes. In general, images of two sources can be resolved if the center of the central maximum of one falls at or beyond the first minimum (dark fringes) of the other. This generally accepted limiting condition for the **resolution** of two diffracted images is known as the **Rayleigh criterion**. The limiting, or minimum, angle of resolution (θ_{min}) for a slit of width w is given by $\theta_{min} = \dfrac{\lambda}{w}$ (where θ_{min} is a pure number and therefore must be expressed in radians). Thus, the images of two sources will be *distinctly* resolved if the angular separation of the sources is greater than λ/w. For *circular* apertures, the minimum angle of resolution is

$\theta_{min} = \dfrac{1.22\,\lambda}{D}$, where D is the diameter of the aperture.

For a microscope, it is convenient to specify the **resolving power**, or the actual separation (s) between two point sources. Because the objects are usually near the focal point of the objective, the minimum distance between two points whose images can be just resolved is $s = f\theta_{min} = \dfrac{1.22\lambda f}{D}$.

Note: The term *resolving power* used here is actually a distance, not power as used in our everyday lives.

Example 25.5   A binary star system in the constellation Orion has an angular separation of 2.5×10^{-5} rad. If the wavelength of the light from the system is $\lambda = 550$ nm, what is the smallest-aperture (diameter) telescope that can just resolve the two stars? (Ignore atmospheric blurring.)

Solution: Given: $\theta_{min} = 2.5 \times 10^{-5}$ rad, $\lambda = 550$ nm $= 550 \times 10^{-9}$ m.
             Find: D.

From $\theta_{min} = \dfrac{1.22\lambda}{D}$, we have $D = \dfrac{1.22\lambda}{\theta_{min}} = \dfrac{1.22(550 \times 10^{-9}\text{ m})}{2.5 \times 10^{-5}\text{ rad}} = 2.7 \times 10^{-2}$ m $= 2.7$ cm.

Due to atmospheric blurring, the diameter of the telescope is much larger than the value predicted here.

Example 25.6   A compound microscope is designed to resolve objects that are 0.010 mm apart. If the focal length of the objective is 4.0 cm and the wavelength of light used is 550 nm, what is the diameter of the aperture of the objective?

Solution: Given: $s = 0.010$ mm $= 0.010 \times 10^{-3}$ m, $f = 4.0$ cm $= 0.040$ m, $\lambda = 550$ nm $= 550 \times 10^{-9}$ m.

Find: D.

The 0.010-mm distance is the resolving power of the microscope.

From $s = \dfrac{1.22\lambda f}{D}$, we have $D = \dfrac{1.22\lambda f}{s} = \dfrac{1.22(550 \times 10^{-9} \text{ m})(0.040 \text{ m})}{0.010 \times 10^{-3} \text{ m}} = 2.7 \times 10^{-3}$ m $= 2.7$ mm.

In reality, the aperture of the microscope must be larger than the predicted value to take other laboratory conditions into account, such as air movement and temperature fluctuations.

---

*5.   Color (Section 25.5)

The **additive primary colors** (additive primaries) are red, blue, and green. The mixing of light of the additive primaries is called the **additive method of color production**. When light of the three additive primaries is mixed in proper portion, the mixture appears white to the eye. Pairs of color combinations that appear white to the eye are called **complementary colors**. The complementary color of red is cyan, of blue is yellow, and of green is magenta, and so forth.

The **subtractive primary pigments** (subtractive primaries) are cyan, yellow, and magenta. A mixture of absorbing pigments results in the subtraction of colors, and one sees the color of light that is not absorbed or subtracted. This is called the **subtractive method of color production.** When the three subtractive primaries are mixed in the proper portion, the mixture appears black (all colors are absorbed) to the eye.

III.   Mathematical Summary

Angular Magnification	$m = \dfrac{\theta}{\theta_0}$ (25.1)	Defines the angular magnification.
Magnification of a Magnifying Glass with Image at Near Point	$m = 1 + \dfrac{25 \text{ cm}}{f}$ (25.3)	Computes the angular magnification of a magnifying glass when the image is at the near point (25 cm).
Magnification of a Magnifying Glass with Image at Infinity	$m = \dfrac{25 \text{ cm}}{f}$ (25.4)	Computes the angular magnification of a magnifying glass when the image is at infinity.

Total Magnification of a Compound Microscope	$M_{total} = M_o m_e = -\dfrac{(25\ cm)L}{f_o f_e}$ (25.5)	Computes the magnification of a compound microscope consisting of an objective and an eyepiece (with $L, f_o,$ and f_e in cm).
Magnification of a Refracting Telescope	$m = -\dfrac{f_o}{f_e}$ (25.6)	Calculates the magnification of a refracting telescope consisting of an objective and an eyepiece.
Minimum Angle of Resolution for a Slit	$\theta_{min} = \dfrac{\lambda}{w}$ (25.7)	Defines the minimum angle of resolution for a slit width w.
Minimum Angle of Resolution for a Circular Aperture	$\theta_{min} = \dfrac{1.22\lambda}{D}$ (25.8)	Defines the minimum angle of resolution for a circular aperture of diameter D.
Resolving Power	$s = f\theta_{min} = \dfrac{1.22\lambda f}{D}$ (25.9)	Defines the resolving power of a circular lens of focal length f.

IV. Solutions of Selected Exercises and Paired/Trio Exercises

5. The pre-flash occurs before the aperture is open and the film exposed. The bright light causes the iris to reduce down (giving a small pupil) so that when the second flash comes momentarily, you don't have a wide opening through which you get the red-eye reflection from the retina.

10. (a) $\boxed{\text{(2) Diverging}}$ contact lens should be prescribed because the person is nearsighted.

(b) The lens is to form an image of an object at infinity at the far point (90 cm).

$d_o = \infty$ and $d_i = -90$ cm (image is on the object side). From the thin-lens equation, $\dfrac{1}{f} = \dfrac{1}{d_o} + \dfrac{1}{d_i}$,

$$P = \frac{1}{f} = \frac{1}{d_o} + \frac{1}{d_i} = \frac{1}{\infty} + \frac{1}{-0.90\ m} = \frac{1}{-0.90\ m} = \boxed{-1.1\ D}.$$

14. (a) The answer is $\boxed{\text{(2) farsightedness}}$. Her near point is farther than normal.

(b) Converging lens will allow her to read the text at the normal near point. The lens is to form an image of an object at 25 cm at the near point (0.80 m).

$d_o = 25$ cm and $d_i = -0.80$ m $= -80$ cm (image on object side).

From the thin-lens equation, $\dfrac{1}{f} = \dfrac{1}{d_o} + \dfrac{1}{d_i}$,

$$\frac{1}{f} = \frac{1}{d_o} + \frac{1}{d_i} = \frac{1}{25\ cm} + \frac{1}{-80\ cm} = 0.0275\ cm^{-1},\ so\ f = \boxed{36\ cm}.$$

18. First calculate the power of the lens from the far points. $d_o = \infty$ and $d_i = -4.0$ m (image on object side).

$$P = \frac{1}{f} = \frac{1}{d_o} + \frac{1}{d_i} = \frac{1}{\infty} + \frac{1}{-4.0 \text{ m}} = -0.25 \text{ D}.$$

For the near points, $d_i = -0.20$ m.

$$\frac{1}{d_o} = P - \frac{1}{d_i} = -0.25 \text{ D} - \frac{1}{-0.20 \text{ m}} = 4.75 \text{ m}^{-1}, \text{ so } d_o = 0.21 \text{ m} = \boxed{21 \text{ cm}}.$$

24. First find his new near point. $d_o = 33$ cm $= 0.33$ m, $P = \frac{1}{f} = +2.0$ D.

$$\frac{1}{d_i} = \frac{1}{f} - \frac{1}{d_o} = +2.0 \text{ D} - \frac{1}{0.33 \text{ m}} = -1.0 \text{ D, so } d_i = -1.0 \text{ m}.$$

Therefore the near point is 1.0 m.

To bring this near point to 25 cm, the power of the new lenses must be

$$P' = \frac{1}{f'} = \frac{1}{0.25 \text{ m}} + \frac{1}{-1.0 \text{ m}} = \boxed{+3.0 \text{ D}}.$$

32. $m = 1 + \dfrac{25 \text{ cm}}{f} = 1 + \dfrac{25 \text{ cm}}{12 \text{ cm}} = \boxed{3.1\times}.$

34. (a) The answer is $\boxed{\text{(1) the near point}}$.

At the near point, $m = 1 + \dfrac{25 \text{ cm}}{f}$; viewing with the relaxed eye, $m = \dfrac{25 \text{ cm}}{f}$.

(b) $m = 1 + \dfrac{25 \text{ cm}}{f} = 1 + \dfrac{25 \text{ cm}}{15 \text{ cm}} = \boxed{2.7\times}$. $m = \dfrac{25 \text{ cm}}{f} = \dfrac{25 \text{ cm}}{15 \text{ cm}} = \boxed{1.7\times}$.

40. $M_{total} = -\dfrac{(25 \text{ cm})L}{f_o f_e} = -\dfrac{(25 \text{ cm})(18 \text{ cm})}{(0.45 \text{ cm})(3.0 \text{ cm})} = -333\times \approx \boxed{-330\times}.$

45. (a) For greater magnification, the focal length of the objective should be as short as possible, and the magnification of the eye piece should be as great as possible. So the answers are:

$\boxed{\text{greatest: 1.6 mm/10}\times\text{; least: 16 mm/5}\times}$.

(b) From the thin-lens equation, $d_o = \dfrac{d_i f_o}{d_i - f_o}$. So $M_o = -\dfrac{d_i}{d_o} = -\dfrac{d_i - f_o}{f_o}$.

Therefore $M_1 = -\dfrac{150 \text{ mm} - 16 \text{ mm}}{16 \text{ mm}} = -8.38\times$, $M_2 = -\dfrac{150 \text{ mm} - 4.0 \text{ mm}}{4.0 \text{ mm}} = -36.5\times$, and

$M_3 = -\dfrac{150 \text{ mm} - 1.6 \text{ mm}}{1.6 \text{ mm}} = -92.8\times.$

Therefore $M_{max} = (-92.8\times)(10\times) = \boxed{-930\times}$ and $M_{min} = (-8.38\times)(5.0\times) = \boxed{-42\times}.$

51. $\boxed{\text{The one with the shorter focal length}}$ should be used as the eyepiece for a telescope. The magnification of the telescope is inversely proportional to the focal length of the eyepiece ($m = -f_o/f_e$).

52. $m = -\dfrac{f_o}{f_e} = -\dfrac{50 \text{ cm}}{2.0 \text{ cm}} = \boxed{-25\times}$. $L = f_o + f_e = 50 \text{ cm} + 2.0 \text{ cm} = \boxed{52 \text{ cm}}$.

56. (a) The answer is $\boxed{\text{(2) increase the physical length of the telescope}}$. The erecting lens will increase the physical length of the telescope. Its purpose is to reverse the orientation of the image formed by the objective so the final image is erect.

(b) With the erecting lens, the image is erect so the magnification is positive.

$m = \dfrac{f_o}{f_e} = \dfrac{40 \text{ cm}}{15 \text{ cm}} = \boxed{2.7\times}$.

(c) $L = f_o + f_e + 4f_i = 40 \text{ cm} + 15 \text{ cm} + 4(20 \text{ cm}) = \boxed{135 \text{ cm}}$.

59. (a) For greater magnification, the focal length of the objective should be as long as possible, and the focal length of the eyepiece should be as short as possible because $m = -\dfrac{f_o}{f_e}$. So the answers are:

maximum magnification: $\boxed{60.0 \text{ cm and } 0.80 \text{ cm}}$; minimum magnification: $\boxed{40.0 \text{ cm and } 0.90 \text{ cm}}$.

(b) $m_1 = -\dfrac{f_{o1}}{f_{e1}} = \dfrac{60.0 \text{ cm}}{0.80 \text{ cm}} = \boxed{-75\times}$, $m_2 = -\dfrac{40.0 \text{ cm}}{0.90 \text{ cm}} = \boxed{-44\times}$.

63. Smaller minimum angle of resolution corresponds to higher resolution because smaller angle of resolution means more details can be resolved.

66. (a) The answer is $\boxed{\text{(2) lower}}$. A wider single slit will have a narrower diffraction pattern so details are not easy to see or the resolution is lower.

(b) According to the Rayleigh criterion, The central maximum of one pattern falls on the first minimum of the other. The angular position of the first minimum is determined by $w \sin \theta \approx w\theta = m\lambda = (1)\,\lambda$ or

$\theta_{min} = \dfrac{\lambda}{w}$.

For the 0.55-mm slit, $\theta_{min} = \dfrac{\lambda}{w} = \dfrac{680 \times 10^{-9} \text{ m}}{0.55 \times 10^{-3} \text{ m}} = \boxed{1.2 \times 10^{-3} \text{ rad}}$.

For the 0.45-mm slit, $\theta_{min} = \dfrac{680 \times 10^{-9} \text{ m}}{0.45 \times 10^{-3} \text{ m}} = \boxed{1.5 \times 10^{-3} \text{ rad}}$.

As expected, the minimum angle of resolution is smaller for the wider single slit.

68. $\theta_{min} = \dfrac{1.22\lambda}{D} = \dfrac{1.22(550 \times 10^{-9} \text{ m})}{8.2 \text{ m}} = \boxed{8.18 \times 10^{-8} \text{ rad}}$.

71. (a) The eye obtain the maximum resolution (smallest minimum angle of resolution) for objects of

$\boxed{\text{(3) blue}}$ color. This is because the minimum angle of resolution is proportional to wavelength,

$\theta_{min} = 1.22\lambda/D$, and blue has the shorter wavelength.

(b) For the 550-nm source, $\theta_{min} = \dfrac{1.22\lambda}{D} = \dfrac{1.22(550 \times 10^{-9}\,\text{m})}{7.0 \times 10^{-3}\,\text{m}} = \boxed{9.6 \times 10^{-5}\,\text{rad}}$.

For the 650-nm source, $\theta_{min} = \dfrac{1.22(650 \times 10^{-9}\,\text{m})}{7.0 \times 10^{-3}\,\text{m}} = \boxed{1.1 \times 10^{-4}\,\text{rad}}$.

As expected, the minimum angle of resolution is smaller (therefore higher resolution) for the source with the short wavelength.

74. (a) Violet light has the highest resolution. The angular separation of the stars must be larger than the minimum angle of resolution.

$\theta_{min} = \dfrac{1.22\lambda}{D} = \dfrac{1.22(400 \times 10^{-9}\,\text{m})}{0.300\,\text{m}} = \boxed{1.63 \times 10^{-6}\,\text{rad}}$.

(b) The lateral distance is $d = \theta L = (1.63 \times 10^{-6}\,\text{rad})(6.00 \times 10^{23}\,\text{m}) = \boxed{9.76 \times 10^{17}\,\text{m}}$.

V. Practice Quiz

1. The farthest distance at which the normal eye can see objects clearly is
   (a) the near point.   (b) the far point.   (c) nearsightedness.   (d) farsightedness.   (e) astigmatism.

2. Which one of the following is not a primary color of light?
   (a) red   (b) yellow   (c) blue   (d) green   (e) both (c) and (d)

3. A nearsighted person wears contact lenses whose lenses have a power of −0.15 D. What is the person's far point?
   (a) 1.5 m   (b) 3.3 m   (c) 6.0 m   (d) 6.7 m   (e) infinity

4. A magnifying glass has a focal length of 5.0 cm. What is the angular magnification if the image is viewed by a relaxed eye?
   (a) 3.0×   (b) 4.0×   (c) 5.0×   (d) 6.0×   (e) 7.0×

5. A compound microscope has an 18-cm barrel and an eyepiece with a focal length of 8.0 mm. What focal length of the objective will give a total magnification of −240×?
   (a) 0.094 cm   (b) 0.13 cm   (c) 1.5 cm   (d) 1.9 cm   (e) 2.3 cm

6. A person is designing a $-10\times$ telescope. If the telescope is limited to a length of 22 cm, what is the approximate focal length of the objective?

(a) 16 cm   (b) 18 cm   (c) 20 cm   (d) 22 cm   (e) 24 cm

7. The 2.4-m (diameter) reflecting Hubble Space Telescope has been placed into Earth orbit by the space shuttle. What angular resolution can this telescope achieve by the Rayleigh criterion if the wavelength is 550 nm?

(a) 5.2×10^{-6} rad   (b) 4.4×10^{-6} rad   (c) 4.6×10^{-7} rad   (d) 2.8×10^{-7} rad   (e) 2.3×10^{-7} rad

8. To decrease the minimum angle of resolution of a microscope,

(a) the diameter of the objective should be decreased.

(b) the diameter of the objective should be increased.

(c) the wavelength of light should be increased.

(d) the microscope's magnification should be more powerful.

(e) none of the above

9. If a farsighted person wears glasses of prescription +3.2D, what is the person's near point if the glasses are very close to the eyes?

(a) 1.25 m   (b) 1.40 m   (c) 1.65 m   (d) 1.75 m   (e) 2.00 m

10. A refracting telescope has an angular magnification m. If the objective focal length is doubled and the eyepiece focal length is halved, what is the new magnification?

(a) $4m$ (b) $2m$ (c) m (d) $m/2$ (e) $m/4$

11. From a spacecraft in orbit 100 km above the Earth's surface, what size features will an astronaut be able to identify with the unaided eye? (Assume a pupil diameter 4.5 mm.)

(a) less than 10 m   (b) greater than 10 m but less than 20 m   (c) greater than 20 m but less than 30 m

(d) greater than 30 m but less than 40 m   (e) greater than 40 m but less than 50 m

12. You are given a convex lens of focal length of 12 cm. What is the maximum magnification when it is used as a simple magnifying glass?

(a) $0.48\times$ (b) $1.48\times$ (c) $2.1\times$ (d) $3.1\times$ (e) $12\times$

Answers to Practice Quiz:

1.b 2.b 3.d 4.c 5.e 6.c 7.d 8.b 9.a 10.a 11.b 12.d

CHAPTER 26

Relativity

I.   Chapter Objectives

Upon completion of this chapter, you should be able to:

1.   summarize the concepts of classical relativity, define inertial and noninertial reference frames, and explain the ether hypothesis and the reasons for its demise.

2.   explain the two postulates of relativity, and how they lead to the relativity of simultaneity.

3.   understand the origins of time dilation and length contraction, and determine the relationship between time intervals and lengths observed in different inertial frames.

4.   understand the relativistically correct expressions for kinetic energy, momentum, and total energy, understand the equivalence of mass and energy, and use the relativistically correct expressions to calculate energy and momentum in elementary particle interactions.

5.   explain the principle of equivalence, and examine some of the predictions of general relativity.

*6.   understand the necessity for a relativistic velocity addition, and investigate relative velocity addition through calculations.

II.   Chapter Summary and Discussion

1.   Classical Relativity and the Michelson-Morley Experiment (Section 26.1)

Relativity is a branch of physics that deals with fast-moving (close to the speed of light, $v \sim c$) particles and observations from two different reference frames. Classical (nonrelativistic) physics is an approximation of relativity at low speeds ($v \ll c$).

An **inertial reference frame** is a nonaccelerating frame in which Newton's first law holds. In an inertial frame, an *isolated* object (one on which there is no net force) is stationary or moves with constant velocity. The **principle of classical (Newtonian) relativity** states: the laws of mechanics are the same in all inertial reference frames.

The speed of light in vacuum was predicted by Maxwell's equations to be 3.00×10^8 m/s, but relative to what reference frame does light have this speed? Classically the speed of light measured from different frames of reference would be expected to differ—possibly to be even greater than 3.00×10^8 m/s if one were approaching the light wave—by simple vector addition of velocities, so it was reasoned that this speed of light that Maxwell predicted *must* be referenced to a particular frame. This conclusion gave rise to the concept of a unique reference frame associated with the proposed medium of transport for electromagnetic waves called the **ether**; an absolute reference frame. As a result, it would seem that Maxwell's equations (which govern light and its speed) did *not* satisfy the Newtonian principle as did the laws of mechanics.

The **Michelson-Morley experiment** was an attempt to detect the ether by using the interference of light in an extremely sensitive **interferometer**. The experiment showed that the speed of light was always *c*, which ruled out the existence of the ether.

2. The Postulates of Special Relativity and the Relativity of Simultaneity
(Section 26.2)

The result of the Michelson-Morley experiment, or the failure to detect the ether, was resolved by the **special theory of relativity** formulated by Albert Einstein. His theory had two basic postulates:

Postulate I (**principle of relativity**):   all the laws of physics (not just mechanics) are the same in all inertial reference frames.

Postulate II (**constancy of the speed of light**):   the speed of light in a vacuum has the same value as in all inertial systems.

The first postulate means that all inertial reference frames are physically equivalent, with physical laws being the same in all of them. It also implies that there are no absolute reference frames.

The second postulate implies that two observers in different inertial reference frames would measure the speed of light to be *c* independent of the speed of the source and/or the observer.

In the language of relativity, an **event** is a "happening." We need to specify the location *and* time to tell an event. Events that are simultaneous in a particular inertial reference frame may *not* be simultaneous when measured in a different inertial frame. Simultaneity is thus a relative, not an absolute, concept. The nonintuitive results of length contraction and time dilation (discussed next) are just two of the many relativistic results that follow directly from the two postulates and the relativity of simultaneity. They are non-intuitive because we are used to dealing with speeds much slower than that of light, for which lengths and time intervals are absolute.

3. The Relativity of Length and Time: Time Dilation and Length Contraction

(Section 26.4)

Two results of the postulates of the special theory of relativity are time dilation and length contraction. **Time dilation** refers to a phenomenon in which a fast-moving clock is measured to run more slowly than a clock at rest in the observer's own frame of reference. Because time intervals are relative, and there are two time intervals, the time measured in a frame at rest with respect to the clock is called the **proper time** (Δt_0). The dilated time (Δt), measured by a clock moving with respect to the frame, and the proper time are related by the equation

$\Delta t = \dfrac{\Delta t_0}{\sqrt{1 - (v/c)^2}}$, where v is the relative speed between the two frames, and c is the speed of light. Many

relativistic equations can be written more simply if we represent the expression $\dfrac{1}{\sqrt{1 - (v/c)^2}}$ as the γ factor:

$\gamma = \dfrac{1}{\sqrt{1 - (v/c)^2}}$. Gamma ($\gamma$) is always greater than or equal to 1. Therefore, the time dilation equation can be

written as $\Delta t = \gamma \Delta t_0$.

Length contraction refers to a phenomenon in which the length of a fast-moving object relative to an observer in an inertial reference frame is less (in the dimension of relative motion) than if the object were at rest in the observer's frame. The length measured in a frame at rest with respect to the object is called the **proper length** (L_0). The contracted length (L) and the proper length are related by $L = \dfrac{L_0}{\gamma} = L_0 \sqrt{1 - (v/c)^2}$.

Note: It is very important to identify the proper time and the proper length in problem solving. They are the time and the length measured in a frame at rest with respect to the clock and the object. Proper time and proper length are also called rest time and rest length.

Time dilation gives rise to another popular relativistic topic—the so called **twin (clock) paradox**. A twin on a space journey relative to the Earth would age more slowly than an Earth-based twin. The twin paradox has been experimentally verified with atomic clocks.

Integrated Example 26.1

One 20-year-old twin brother takes a space trip with a speed of $0.70c$ for 30 years according to a clock on the spaceship. The other twin remains on the Earth. (a) On returning to the Earth, the traveling twin is (1) younger, (2) the same age as, or (3) older than the Earth-based twin. Explain. (b) What are the ages of the two twins once the traveling twin returns to the Earth.

(a) Conceptual Reasoning:

Since the time interval of 30 years is measured by the traveling twin with his own clock, it is the proper time. This same 30 years will be diluted for the Earth-bound twin so it will be more than 30 years for him. Therefore the traveling twin will be (1) younger than the Earth-based twin.

(b) Quantitative Reasoning and Solution:

Given: $\Delta t_o = 30$ years, $v = 0.70c$.

Find: Ages of the traveling twin and the Earth-based twin.

The proper time is 30 years according to the traveling twin.

The age of the traveling twin is simply 20 years + 30 years = 50 years.

According to the Earth-based twin, 30 years on the spaceship is diluted to

$$\Delta t = \gamma \Delta t_o = \frac{\Delta t_o}{\sqrt{1 - (v/c)^2}} = \frac{(30 \text{ years})}{\sqrt{1 - (0.70)^2}} = 42 \text{ years on the Earth.}$$

Therefore, the age of the Earth-based twin is 20 years + 42 years = 62 years.

As expected, the traveling twin is younger than the Earth-based twin when he returns from space travel.

Example 26.2 A spaceship is moving toward you with a speed of 0.75c. By what percentage will the spaceship's length change compared with its length of 15 m when it is at rest?

Solution: Given: $L_o = 15$ m, $v = 0.75c$.

Find: L/L_o.

The proper length of the spaceship is 15 m.

From length contraction, $L = \dfrac{L_o}{\gamma} = L_o \sqrt{1 - (v/c)^2} = (15 \text{ m}) \sqrt{1 - (0.75)^2} = 9.9$ m.

Thus, the spaceship appears to be $\dfrac{9.9 \text{ m}}{15 \text{ m}} = 0.66 = 66\%$ as compared with its length at rest. That means the length decreases by 34%.

Example 26.3 The closest star to our solar system is Alpha Centauri, which is 4.30 light years away. A spaceship with a constant velocity of 0.800c relative to the Earth travels toward the star.

(a) What distance does the spaceship travel according to a passenger on the ship?

(b) How much time would elapse on a clock on board the spaceship?

(c) How much time would elapse on a clock on the Earth?

Solution: Given: $L_o = 4.30$ light years, $v = 0.800c$.

Find: (a) L (b) Δt_o (c) Δt.

The 4.30 light-year distance is measured by observers on the Earth, so it is the proper length. The proper time is the time measured with a clock on the spaceship.

(a) From length contraction,

$$L = \frac{L_o}{\gamma} = L_o \sqrt{1 - (v/c)^2} = (4.30 \text{ light years}) \sqrt{1 - (0.800)^2} = 2.58 \text{ light years}.$$

(b) The proper time (the time measured on the spaceship) is

$$\Delta t_o = \frac{L}{v} = \frac{2.58 \text{ light years}}{0.800c} = \frac{2.58c \text{ years}}{0.800c} = 3.23 \text{ years}.$$

(c) The dilated time (the time measured by an Earth-based clock) is then

$$\Delta t = \gamma \Delta t_o = \frac{\Delta t_o}{\sqrt{1 - (v/c)^2}} = \frac{3.23 \text{ years}}{\sqrt{1 - (0.800)^2}} = 5.38 \text{ years}.$$

Or $\Delta t = \dfrac{L_o}{v} = \dfrac{4.30 \text{ light years}}{0.800c} = \dfrac{4.30c \text{ years}}{0.800c} = 5.38 \text{ years}.$

The same result is obtained two different ways.

---

4. Relativistic Kinetic Energy, Momentum, Total Energy, and Mass-Energy Equivalence (Section 26.4)

Other important quantities in the special theory of relativity are relativistic kinetic energy, momentum, total energy, and mass–energy equivalence.

Relativistic kinetic energy is given by $K = \left[\dfrac{1}{\sqrt{1 - (v/c)^2}} - 1 \right] mc^2 = (\gamma - 1)mc^2$.

Relativistic momentum is given by $\vec{p} = \dfrac{m\vec{v}}{\sqrt{1 - (v/c)^2}} = \gamma m \vec{v}$.

Total relativistic energy is equal to $E = \dfrac{mc^2}{\sqrt{1 - (v/c)^2}} = \gamma mc^2$. When $v = 0$, $E = E_o = mc^2$, which is called the **rest energy**. E can also be written as $E = K + E_o = K + mc^2 = \gamma E_o$. The expression $E_o = mc^2$ is the famous **mass-energy equivalence** because it points out that mass is also a form of energy. In particular, a particle has rest energy E_o associated with its mass m. For speeds below 10% of the speed of light, the use of the nonrelativistic formulas ($K = \frac{1}{2}mv^2$, etc.) is acceptable, since it will cause errors of less than 1% error.

Example 26.4   The kinetic energy of a proton is 80% of its total energy.

(a) What is the speed of the proton?

(b) What is the magnitude of the momentum of the proton?

(c) What is the total energy of the proton?

Solution: Given: $K = 0.80E$, $m = 1.67 \times 10^{-27}$ kg.

Find: (a) v (b) p (c) E.

(a) Because the total energy is the sum of the kinetic energy and the rest energy,

$$E = K + E_o = 0.80E + E_o, \; E_o = 0.20E.$$

Also from $E = \gamma E_o$, $\gamma = \dfrac{E}{E_o} = \dfrac{1}{0.20} = 5.0$.

Because $\dfrac{1}{\gamma} = \sqrt{1 - (v/c)^2} = 0.20$, $v = c\sqrt{1 - (0.20)^2} = 0.98c$.

(b) $p = \gamma mv = (5.0)(1.67 \times 10^{-27}$ kg$)(0.98)(3.00 \times 10^8$ m/s$) = 2.5 \times 10^{-18}$ kg·m/s.

(c) $E = \gamma E_o = \gamma mc^2 = (5.0)(1.67 \times 10^{-27}$ kg$)(3.00 \times 10^8$ m/s$)^2 = 7.5 \times 10^{-10}$ J $= 4.7$ GeV.

5.   The General Theory of Relativity (Section 26.5)

The special theory of relativity applies only to inertial reference frames, *not* to noninertial (accelerating) systems. The **general theory of relativity** considers accelerating frames and is essentially a gravitational theory. Its **principle of equivalence** states that an inertial reference frame in a uniform gravitational field is equivalent to a reference frame in the absence of a gravitational field that has a constant acceleration with respect to that inertial frame. It basically means that no experiment performed in a closed system can distinguish between the effects of a gravitational field and the effects of an acceleration.

Some of the predictions of the general theory of relativity are gravitational light bending, gravitational lensing, black holes, and the gravitational redshift. One way a **black hole** can form is from the gravitational collapse of a massive neutron star. Such an object has a density so great and a gravitational field so intense that nothing (including light) can escape it. The critical radius around a black hole from which light cannot escape is given by the **Schwarzschild radius**, $R = \dfrac{2GM}{c^2}$, where G is the universal gravitational constant, and M is the mass within the radius. The boundary of a sphere of radius R defines what is called the **event horizon**. Any event occurring within this horizon is invisible to an observer outside, since light cannot escape.

Example 26.5   What radius would our Sun have to have in order for light not to be able to escape from it?

Solution: Given: $M = 2.0 \times 10^{30}$ kg, $G = 6.67 \times 10^{-11}$ N·m²/kg².   Find: R.

For light not to be able to escape, the Sun would have to be a black hole. The Schwarzschild radius would

be $R = \dfrac{2GM}{c^2} = \dfrac{2(6.67 \times 10^{-11} \text{ N·m}^2/\text{kg}^2)(2.0 \times 10^{30} \text{ kg})}{(3.00 \times 10^8 \text{ m/s})^2} = 3.0 \times 10^3$ m = 3.0 km.

*6.   Relativistic Velocity Addition (Section 26.6)

In situations dealing with relative velocities for fast-moving particles, the relativistic velocity addition must

be used, $u = \dfrac{v + u'}{1 + \dfrac{vu'}{c^2}}$, where

v is the velocity of object 1 with respect to an inertial observer (e.g., on the Earth),

u' is the velocity of object 2 with respect to object 1, and

u is the velocity of object 2 with respect to an inertial observer (e.g., on the Earth).

Example 26.6   A spaceship moves away from the Earth with a speed of $0.80c$. The spaceship then fires a missile with a speed of $0.50c$ relative to the ship. What is the velocity of the missile measured by observers on the Earth if

(a) the missile is fired away from the Earth?

(b) the missile is fired toward the Earth?

Solution: Given: $v = 0.80c$,   (a) $u' = +0.50c$,   (b) $u' = -0.50c$.

Find: u (a) and (b).

Here we assume away from the Earth as the positive direction for velocity.

(a) $u = \dfrac{v + u'}{1 + \dfrac{vu'}{c^2}} = \dfrac{0.80c + 0.50c}{1 + (0.80)(0.50)} = 0.93c$ away from the Earth.

According to classical theory, $u = v + u' = 0.80c + 0.50c = 1.3c$!

(b) $u = \dfrac{0.80c + (-0.50c)}{1 + (0.80)(-0.50)} = 0.50c$ away from the Earth.

According to the classical theory, $u = 0.80c + (-0.50c) = 0.30c$!

III. Mathematical Summary

Time Dilation	$\Delta t = \dfrac{\Delta t_0}{\sqrt{1-(v/c)^2}}$ (26.3)	Computes the dilated time in terms of proper time.
γ (dimensionless relativistic factor)	$\gamma = \dfrac{1}{\sqrt{1-(v/c)^2}}$ (26.4)	Defines the dimensionless relativistic factor.
Time Dilation Expressed with γ	$\Delta t = \gamma \Delta t_0$ (26.5)	Computes the dilated time in terms of γ and proper time.
Length Contraction	$L = \dfrac{L_0}{\gamma} = L_0\sqrt{1-(v/c)^2}$ (26.7)	Computes the contracted length in terms of proper length.
Relativistic Kinetic Energy	$K = \left[\dfrac{1}{\sqrt{1-(v/c)^2}} - 1\right]mc^2$ $= (\gamma - 1)mc^2$ (26.8)	Defines relativistic kinetic energy.
Relativistic Momentum	$\vec{p} = \dfrac{m\vec{v}}{\sqrt{1-(v/c)^2}}$ $= \gamma m \vec{v}$ (26.9)	Defines relativistic momentum.
Relativistic Total Energy	$E = \dfrac{mc^2}{\sqrt{1-(v/c)^2}}$ $= \gamma mc^2$ (26.10)	Computes total relativistic energy.
Rest Energy	$E_0 = mc^2$ (26.11)	Defines rest energy.
Relativistic Total Energy	$E = K + E_0 = K + mc^2$ (26.12)	Computes total relativistic energy.

IV. Solutions of Selected Exercises and Paired Exercises

8. (a) The velocity relative to ground is 200 km/h $+ (-35$ km/h$) = \boxed{165 \text{ km/h}}$.

   (b) The velocity relative to ground is 200 km $+ 25$ km/h $= \boxed{225 \text{ km/h}}$.

10.   (a) The time it takes is $\boxed{\text{(1) longer}}$. Although it takes less time on the trip in the direction of the current, it takes more time on the trip in the direction opposite the current. The extra time in the opposite direction more than offsets the lesser time in the direction of the current.

(b) When there is no current, the time is $t_1 = \dfrac{1000 \text{ m}}{20 \text{ m/s}} + \dfrac{1000 \text{ m}}{20 \text{ m/s}} = 100 \text{ s} = \boxed{1.7 \text{ min}}$.

In the direction of current: the relative velocity is $20 \text{ m/s} + 5.0 \text{ m/s} = 25 \text{ m/s}$.

In the direction opposite the current: the relative velocity is $20 \text{ m/s} - 5.0 \text{ m/s} = 15 \text{ m/s}$.

So the time is $t_2 = \dfrac{1000 \text{ m}}{25 \text{ m/s}} + \dfrac{1000 \text{ m}}{15 \text{ m/s}} = 107 \text{ s} = \boxed{1.8 \text{ min}}$.

16.   No, it is not possible. Any inertial observer cannot measure an object with mass to travel at a speed equal to or greater than the speed of light, c.

24.   (a) You are measuring the proper time because you and your clock are in the same frame of reference (no relative motion).

(b) Your professor measures the proper length of the spacecraft because your professor and the spacecraft are in the same frame of reference (no relative motion).

26.   (a) According to your professor, your pulse rate is $\boxed{\text{(3) less than 80 beats/min}}$. Due to time dilation, the time it takes for the heart to beat will increase, which will decrease the pulse rate.

(b) The proper time for one beat is $\dfrac{1}{80 \text{ beats/min}} = \dfrac{1}{80} \text{ min/beat}$.

So $\Delta t = \dfrac{\Delta t_o}{\sqrt{1 - v^2/c^2}} = \dfrac{1/80 \text{ min/beat}}{\sqrt{1 - 0.85^2}} = \dfrac{1}{42} \text{ min/beat}$.

Therefore, the number of beats per min is $\boxed{42 \text{ beats/min}}$.

28.   (a) The length of the field will, according to the astronaut, be $\boxed{\text{(3) shorter than 100 m}}$ due to length contraction.

(b) The proper length is 100 m. $L = L_o \sqrt{1 - v^2/c^2} = (100 \text{ m}) \sqrt{1 - 0.75^2} = \boxed{66 \text{ m}}$.

(c) $\boxed{100 \text{ m}}$ is the proper length.

30.   Earth twin: 25 years + 39 years = $\boxed{64 \text{ years old}}$.

Find the proper time for the one on the spaceship because 39 years is measured according to Earth time.

$\Delta t = \dfrac{\Delta t_{\text{o}}}{\sqrt{1 - v^2/c^2}}$, so $\Delta t_{\text{o}} = \Delta t \sqrt{1 - v^2/c^2} = (39 \text{ years}) \sqrt{1 - 0.95^2} = 12.2 \text{ years}$.

Traveling twin: 25 years + 12.2 years = $\boxed{37 \text{ years old}}$.

34.   (a) The altitude is still 15.0 m because it is perpendicular to the relative velocity. The proper length of the base is 40.0 m. So the base is $L = L_{\text{o}} \sqrt{1 - v^2/c^2} = (40.0 \text{ m}) \sqrt{1 - 0.90^2} = 17.44 \text{ m}$.

Therefore, the area is $A = \frac{1}{2}(17.44 \text{ m})(15.0 \text{ m}) = \boxed{131 \text{ m}^2}$.

(b) The angle between the hypotenuse and the base is $\theta = \tan^{-1}\left(\dfrac{15.0}{17.44}\right) = \boxed{40.7°}$.

38.   (a) The only thing that matters is that $\boxed{\text{(3) they are moving relative to one another}}$ because velocity is relative.

(b) The length of the meterstick is the proper length, and $\dfrac{v}{c} = \dfrac{1.0 \text{ yd}}{1.0 \text{ m}} = \dfrac{3(0.3048) \text{ m}}{1.0 \text{ m}} = 0.9144$.

$L = L_{\text{o}} \sqrt{1 - v^2/c^2}$, so $v = c \sqrt{1 - L^2/L_{\text{o}}^2} = c \sqrt{1 - 0.9144^2} = \boxed{0.40c}$.

46. $E_{\text{o}} = mc^2 = (9.11 \times 10^{-31} \text{ kg})(3.00 \times 10^8 \text{ m/s})^2 = 8.20 \times 10^{-14} \text{ J} \approx 0.511 \text{ MeV}$.

$E = \gamma E_{\text{o}} = \dfrac{1}{\sqrt{1 - v^2/c^2}} E_{\text{o}} = \dfrac{1}{\sqrt{1 - 0.600^2}}(0.511 \text{ MeV}) = \boxed{0.639 \text{ MeV}}$.

52. $E_{\text{o}} = mc^2 = (9.11 \times 10^{-31} \text{ kg})(3.00 \times 10^8 \text{ m/s})^2 = 8.20 \times 10^{-14} \text{ J} \approx 0.511 \text{ MeV}$.

From $E = \gamma E_{\text{o}} = \dfrac{E_{\text{o}}}{\sqrt{1 - v^2/c^2}}$, we have $v = c \sqrt{1 - E_{\text{o}}^2/E^2} = c \sqrt{1 - (0.511)^2/(2.8)^2} = 0.983c$.

So $p = \dfrac{mv}{\sqrt{1 - v^2/c^2}} = \dfrac{(9.11 \times 10^{-31} \text{ kg})(0.983)(3.00 \times 10^8 \text{ m/s})}{\sqrt{1 - 0.983^2}} = \boxed{1.5 \times 10^{-21} \text{ kg·m/s}}$.

56.   (a) The Sun's mass will $\boxed{\text{(3) decrease}}$ because some of it is converting to energy.

(b) $E_{\text{o}} = mc^2 = (1.989 \times 10^{30} \text{ kg})(3.00 \times 10^8 \text{ m/s})^2 = 1.790 \times 10^{47} \text{ J}$.

Since power is $P = \dfrac{E_{\text{o}}}{t}$, we have $t = \dfrac{E_{\text{o}}}{P} = \dfrac{1.790 \times 10^{47} \text{ J}}{3.827 \times 10^{26} \text{ W}} = 4.678 \times 10^{20} \text{ s} = \boxed{1.483 \times 10^{13} \text{ years}}$.

(c) This tells us that $\boxed{\text{nowhere near 100\% of the Sun's mass is converted into light}}$.

58.   (a) Water will have $\boxed{\text{(1) more}}$ mass than ice because energy is needed to convert ice to water, and that energy becomes part of the internal energy of the water.

(b) The energy required to convert ice to water is $\Delta E = mL_f = (1.0 \text{ kg})(3.33 \times 10^5 \text{ J/kg}) = 3.33 \text{ J}$.

From $E_o = mc^2$, we have $\Delta E_o = \Delta mc^2$. So $\Delta m = \dfrac{E}{c^2} = \dfrac{3.33 \times 10^5 \text{ J}}{(3.00 \times 10^8 \text{ m/s})^2} = \boxed{3.7 \times 10^{-12} \text{ kg}}$.

$\boxed{\text{No}}$, this is not detectable as it is extremely small.

68. (a) The event horizon for Jupiter would be $\boxed{\text{(1) larger than}}$ than the Earth's because Jupiter has more mass

and the radius of the event horizon is directly proportion to the mass, $R = \dfrac{2GM}{c^2}$.

(b) $R_E = \dfrac{2GM_E}{c^2} = \dfrac{2(6.67 \times 10^{11} \text{ N·m}^2/\text{kg}^2)(6.0 \times 10^{24} \text{ kg})}{(3.00 \times 10^8 \text{ m/s})^2} = 8.9 \times 10^{-3} \text{ m} = \boxed{8.9 \text{ mm}}$.

$R_J = \dfrac{2GM_J}{c^2} = \dfrac{2G(318M_E)}{c^2} = 318R_E = \boxed{2.8 \text{ m}}$.

72. $u = \dfrac{v + u'}{1 + vu'/c^2} = \dfrac{0.40c + (-0.15c)}{1 + (0.40)(-0.15)} = \boxed{0.27c, \text{ same direction as spacecraft}}$.

74. (a) The speed of one ship relative to the other is $\boxed{\text{(3) less than } c}$ because c is the upper limit for all speeds.

(b) $u = \dfrac{v + u'}{1 + vu'/c^2} = \dfrac{0.60c + 0.60c}{1 + (0.60c)(0.60c)/c^2} = \boxed{0.88c}$.

76. (a) The answer is $\boxed{\text{(1) the astronaut in the ship}}$.

(b) From $\Delta t = \dfrac{\Delta t_o}{\sqrt{1 - v^2/c^2}}$, we have $\Delta t_o = \sqrt{1 - v^2/c^2}\, \Delta t$.

So the time difference is $\Delta t - \Delta t_o = \left(1 - \sqrt{1 - v^2/c^2}\right)t = \left(1 - \sqrt{1 - 0.60^2}\right)(24 \text{ h}) = \boxed{4.8 \text{ h}}$.

(c) Also since $L = L_o\sqrt{1 - v^2/c^2}$, $L_o = \dfrac{L}{\sqrt{1 - v^2/c^2}} = \dfrac{110 \text{ m}}{\sqrt{1 - 0.60^2}} = \boxed{1.4 \times 10^2 \text{ m}}$.

78. (a) The answer is $\boxed{\text{(3) only length}}$.

(b) $L = L_o\sqrt{1 - v^2/c^2} = (50 \text{ m})\sqrt{1 - 0.65^2} = 38 \text{ m}$. The height and width are the same.

So the dimensions are $\boxed{\text{length 38 m; height 2.5 m; width 2.0 m}}$.

84. (a) The proper time is in reference C.

$\Delta t_B = \dfrac{\Delta t_o}{\sqrt{1 - v^2/c^2}} = \dfrac{1.00 \text{ h}}{\sqrt{1 - 0.50^2}} = 1.1547 \text{ h} = \boxed{1.15 \text{ h}}$.

(b) The relative speed between reference A and C is

$$u = \frac{v + u'}{1 + v u'/c^2} = \frac{0.50c + 0.90c}{1 + (0.50c)(0.90c)/c^2} = 0.9655c.$$

$$\Delta t_A = \frac{\Delta t_o}{\sqrt{1 - v^2/c^2}} = \frac{1.00 \text{ h}}{\sqrt{1 - 0.9655^2}} = \boxed{3.84 \text{ h}}.$$

(c) The proper length is in reference A.

$$L_B = L_o \sqrt{1 - v^2/c^2} = (10 \text{ m}) \sqrt{1 - 0.90^2} = 4.359 \text{ m} = \boxed{4.36 \text{ m}}.$$

(d) $L_C = L_o \sqrt{1 - v^2/c^2} = (10 \text{ m}) \sqrt{1 - 0.9655^2} = \boxed{2.60 \text{ m}}$.

V. Practice Quiz

1. A spaceship is traveling with a speed of $0.80c$ relative to an observer. How fast would light travel from the headlights of the ship relative to that observer?

    (a) $0.20c$ (b) $0.80c$ (c) c (d) $0.90c$ (e) $1.8c$

2. What is the momentum in kg·m/s of an electron when it is moving with a speed of $0.75c$?

    (a) 1.0×10^{-22} (b) 2.0×10^{-22} (c) 2.6×10^{-22} (d) 3.1×10^{-22} (e) 4.6×10^{-22}

3. As the speed of a particle approaches the speed of light the rest energy of the particle

    (a) increases.   (b) decreases.   (c) remains the same.   (d) approaches zero.   (e) none of the preceding

4. A spaceship takes a nonstop journey to a planet and returns in 10 h according to a clock on the spaceship. If the speed of the spaceship is $0.80c$, how much time has elapsed on the Earth?

    (a) 6.0 h   (b) 6.3 h   (c) 10 h   (d) 15 h   (e) 17 h

5. The length of a spaceship is 10 m when it is at rest. If the spaceship travels past you with its velocity $(0.70c)$ parallel to its length, what length does it appear to you?

    (a) 5.5 m   (b) 7.1 m   (c) 10 m   (d) 14 m   (e) 18 m

6. What is the total energy of an electron moving with a speed of $0.95c$?

    (a) 2.6×10^{-13} J   (b) 8.2×10^{-14} J   (c) 1.1×10^{-13} J   (d) 1.2×10^{-14} J   (e) 3.7×10^{-14} J

7. What was the significance of the result of the Michelson-Morley experiment?

    (a) verified the existence of the ether   (b) proved that the speed of light is a constant

    (c) showed length contraction   (d) showed time dilation

(e) verified gravitational light bending

8.   How much fuel mass does a nuclear power plant that is capable of delivering 3.0 GW (3.0×10^9 W) lose in one hour?

(a) 1.2 kg   (b) 0.12 kg   (c) 12 g   (d) 1.2 g   (e) 0.12 g

9.   What is the speed of a proton if its total energy is twice its rest energy?

(a) 0.866c (b) 0.750c (c) 0.707c (d) 0.581c (e) 0.500c

10.   A space traveler in a ship moves away from the Earth with a speed of 0.30c when a missile is fired with a speed of 0.80c back toward the Earth relative to the ship. How fast does the missile appear to travel toward the Earth relative to the Earth?

(a) 0.40c (b) 0.50c (c) 0.66c (d) 0.89c (e) c

11.   The Schwarzschild radius is the radius of

(a) a black hole.   (b) our Sun.   (c) a star.   (d) an event horizon.   (e) none of the preceding.

12.   How fast must a meterstick be moving, parallel to its length, relative to a stationary observer so that he measures its length to be 0.50 m?

(a) 0.50c (b) 0.71c (c) 0.75c (d) 0.87c (e) 1.0c

Answers to Practice Quiz:

1.c 2.d 3.c 4.e 5.b 6.a 7.b 8.e 9.a 10.c 11.d 12.d

CHAPTER 27

I.   Chapter Objectives

Upon completion of this chapter, you should be able to:

1.   define blackbody radiation and use Wien's law, and understand how Planck's hypothesis paved the way for quantum ideas.

2.   describe the photoelectric effect, explain how it can be understood by assuming that light energy is carried by particles, and summarize the properties of photons.

3.   understand how the photon model of light explains the scattering of light from electrons (the Compton effect), and calculate the wavelength of the scattered light in the Compton effect.

4.   understand how the Bohr model of the hydrogen atom explains atom's emission and absorption spectra, calculate the energies and wavelengths of emitted and absorbed photons for transitions in atomic hydrogen, and understand how the generalized concept of atomic energy levels can explain other atomic phenomena.

5.   understand some of the practical applications of the quantum hypothesis—in particular, the laser.

II.   Chapter Summary and Discussion

1.   Quantization: Planck's Hypothesis (Section 27.1)

One of the problems scientists had at the end of the 19th century was how to explain **thermal radiation**, the continuous spectra of radiation emitted by hot objects. A **blackbody** is an ideal system that absorbs and emits all radiation that falls on it. The wavelength of maximum radiation (spectral component), λ_{max}, is inversely proportional to its absolute temperature, T, and obeys Wien's displacement law, $\lambda_{max} T = 2.90 \times 10^{-3}$ m·K.

Classically, the intensity of the blackbody radiation at a particular wavelength is predicted to be proportional to $\frac{1}{\lambda^4}$

(Chapter 11), which leads to what is sometimes called the *ultraviolet catastrophe*—"ultraviolet" because disagreement between theory and experiment occurs for short wavelengths beyond the violet end of the visible spectrum, and "catastrophe" because it predicts that the emitted intensity at these short wavelengths will become infinitely large.

Note: Wien's displacement law relates the wavelength of the *most intense* radiation and the absolute temperature (Kelvin), not maximum wavelength and temperature. λ_{max} does not mean it is the maximum or longest wavelength emitted.

Max Planck successfully explained the spectrum of blackbody radiation by proposing a radical hypothesis. According to **Planck's hypothesis**, the energy of a thermal oscillator is *quantized*. That is, an oscillator can have only *discrete*, or particular, amounts of energy, rather than a continuous distribution of energies. The smallest quantum amount of energy is given by $E = hf$, where $h = 6.63 \times 10^{-34}$ J·s is called **Planck's constant**. This amount is called a **quantum** of energy. The thermal oscillators in a blackbody can have only integer multiples of this quantum of energy, $E_n = n(hf)$.

Example 27.1   What is the most intense color of light emitted by a giant star of surface temperature 4400 K? What is the color of the star?

Solution: Given: $T = 4400$ K.

   Find: λ_{max}.

Wien's displacement law, $\lambda_{max} T = 2.90 \times 10^{-3}$ m·K, gives the most intense wavelength or color of light.

$$\lambda_{max} = \frac{2.90 \times 10^{-3} \text{ m·K}}{T} = \frac{2.90 \times 10^{-3} \text{ m·K}}{4400 \text{ K}} = 659 \text{ nm}.$$

This is in the red end of the visible spectrum. This star actually emits all colors of light but red is the most intense so the color of the star is red.

---

2.   Quanta of Light: Photons and the Photoelectric Effect (Section 27.2)

A quantum or packet of light energy is referred to as a **photon**, and each photon has an energy $E = hf$, where f is the frequency of light. This description suggests that light may sometimes behave as discrete quanta, or "particles" of energy, rather than as a wave.

Einstein used the photon theory to explain the **photoelectric effect**, another area in which the classical (wave) description of light was inadequate. Some materials are *photosensitive*, that is, when light strikes their surface, electrons are emitted and a current may be established; however, if the frequency of light is below a certain cutoff value, *no photoelectrons* are emitted no matter how strong the light intensity is and how long light is incident on the material! If the frequency is higher than the certain cutoff value, photoelectrons are emitted instantaneously, no matter how low the light intensity is. The intensity of light apparently has to do only with the number of photoelectrons. Classical wave physics cannot explain these observations.

To study the maximum kinetic energy K_{max} of the photoelectrons, a **stopping potential** V_0 is applied to a beam of electrons. When there is no photocurrent, the maximum kinetic energy is related to the stopping potential by $K_{max} = eV_0$, where eV_0 is the work needed to stop the most energetic photoelectrons. The minimum amount of energy needed to free the electrons from the material is called the **work function** (ϕ_0). According to energy conservation, $hf = K_{max} + \phi_0$; that is, the energy of the absorbed photon goes into the work of freeing the electron, and the rest is carried off by that emitted electron as kinetic energy. The **threshold frequency**, the lowest frequency of light that can release photoelectrons, corresponds to photoelectrons having zero kinetic energy, so

$$hf_0 = 0 + \phi_0, \text{ or } f_0 = \frac{\phi_0}{h}.$$

Note: We are often given the wavelength of light (in nm) rather than the frequency and energy. The energy of the photon in units of eV can be calculated quickly from $E = \dfrac{1.24 \times 10^3 \text{ eV} \cdot \text{nm}}{\lambda}$.

Example 27.2 What is the photon energy of visible light having a wavelength of 632.8 nm?

Solution: Given: $\lambda = 632.8 \text{ nm} = 632.8 \times 10^{-9} \text{ m}$. Find: E.

The frequency of light is calculated from $f = \dfrac{c}{\lambda} = \dfrac{3.00 \times 10^8 \text{ m/s}}{632.8 \times 10^{-9} \text{ m}} = 4.74 \times 10^{14} \text{ Hz}$.

Then the energy of a photon is $E = hf = (6.63 \times 10^{-34} \text{ J} \cdot \text{s})(4.74 \times 10^{14} \text{ Hz}) = 3.14 \times 10^{-19} \text{ J} = 1.96 \text{ eV}$.

Or we can use $E = \dfrac{1.24 \times 10^3 \text{ eV} \cdot \text{nm}}{\lambda} = \dfrac{1.24 \times 10^3 \text{ eV} \cdot \text{nm}}{632.8 \text{ nm}} = 1.96 \text{ eV}$.

Example 27.3 A metal has a work function of 4.5 eV. Find the maximum kinetic energy of the emitted photoelectrons if the wavelength of light falling on the metal is (a) 300 nm (b) 250 nm.

Solution: Given: $\phi_0 = 4.5 \text{ eV}$. (a) $\lambda = 300 \text{ nm}$ (b) $\lambda = 250 \text{ nm}$.

Find: K_{max} in (a) and (b).

(a) From energy conservation, $E = hf = K_{max} + \phi_0$, we have

$$K_{max} = E - \phi_0 = \frac{1.24 \times 10^3 \text{ eV} \cdot \text{nm}}{\lambda} - \phi_0 = \frac{1.24 \times 10^3 \text{ eV} \cdot \text{nm}}{300 \text{ nm}} - 4.5 \text{ eV} = -0.37 \text{ eV}.$$

Because kinetic energy cannot be negative, this result indicates that no photoelectrons are ejected with the 300 nm light.

(b) $K_{max} = \dfrac{1.24 \times 10^3 \text{ eV} \cdot \text{nm}}{250 \text{ nm}} - 4.5 \text{ eV} = 0.46 \text{ eV}$. There is photoelectrons with the 250 nm light.

Example 27.4   When light of wavelength 350 nm is incident on a metal surface, the stopping potential of the photoelectrons is measured to be 0.500 V.

(a) What is the work function of the metal?

(b) What is the threshold frequency of the metal?

(c) What is the maximum kinetic energy of the photoelectrons?

Solution: Given: $\lambda = 350$ nm, $V_0 = 0.500$ V or $eV_0 = 0.500$ eV

Find: (a) ϕ_0 (b) f_0 (c) K_{max}.

(a) According to energy conservation, $E = hf = K_{max} + \phi_0 = eV_0 + \phi_0$, we have

$$\phi_0 = E - eV_0 = \frac{1.24 \times 10^3 \text{ eV·nm}}{\lambda} - eV_0 = \frac{1.24 \times 10^3 \text{ eV·nm}}{350 \text{ nm}} - 0.500 \text{ eV} = 3.04 \text{ eV}.$$

(b) $f_0 = \dfrac{\phi_0}{h} = \dfrac{(3.04 \text{ eV})(1.60 \times 10^{-19} \text{ J/eV})}{6.63 \times 10^{-34} \text{ J·s}} = 7.34 \times 10^{14}$ Hz.

(c) The maximum kinetic energy is equal to the stopping potential, eV_0, expressed in eV.

$K_{max} = eV_0 = 0.500$ eV.

3.   Quantum "Particles": The Compton Effect (Section 27.3)

The **Compton effect** is the increase in wavelength of light scattered by electrons or other charged particles. When a photon collides with an electron, a scattered photon of longer wavelength (lower frequency or lower energy because the electron carries off some energy) emerges. From the conservation of relativistic momentum and total energy, the shift is given by $\Delta \lambda = \lambda - \lambda_0 = \lambda_C (1 - \cos \theta)$, where $\lambda_C = h/(mc) = 2.43 \times 10^{-12}$ m $= 2.43 \times 10^{-3}$ nm is called the *Compton wavelength* of the electron. Because the Compton shift is very small, it is significant only for X-ray and gamma-ray scattering where the wavelengths are on the order of λ_C.

Note: The Compton wavelength is *not* the wavelength of the electron, it is the characteristic wavelength shift that reoccurs when photons are scattered by an electron. If light is scattered by a proton, the Compton wavelength will be that of the proton and will be much shorter than that of the electrons. (Why?)

Einstein's and Compton's successes in explaining electromagnetic phenomena in terms of quanta left scientists with two apparently competing theories of electromagnetic radiation: the wave theory and the photon theory. The two theories gave rise to a description that is called the **dual nature of light**; that is, light apparently behaves sometimes as a wave and at other times as photons or "particles."

Example 27.5   X-rays of wavelength 0.200 nm are scattered by a metal. The wavelength shift is observed to be 1.50×10^{-12} m at a certain scattering angle measured relative to the incoming X-ray.

          (a) What is the scattering angle?

          (b) What is the maximum shift possible for the Compton efffect?

Solution: Given: $\lambda = 0.200$ nm $= 2.00 \times 10^{-10}$ m, $\Delta\lambda = 1.50 \times 10^{-12}$ m.

               Find: (a) θ (b) $\Delta\lambda_{max}$.

(a) From $\Delta\lambda = \lambda_C(1 - \cos\theta) = (2.43 \times 10^{-12}$ m$)(1 - \cos\theta)$, we have

$$\cos\theta = 1 - \frac{\Delta\lambda}{\lambda_C} = 1 - \frac{1.50 \times 10^{-12}\text{ m}}{2.43 \times 10^{-12}\text{ m}} = 0.617. \text{ So } \theta = \cos^{-1}(0.617) = 51.9°.$$

(b) Maximum shift occurs when $\theta = 180°$ (or $\cos\theta = -1$).

$$\Delta\lambda_{max} = \lambda_C (1 - \cos 180°) = 2\lambda_C = 4.86 \times 10^{-12}\text{ m.}$$

4.   The Bohr Theory of the Hydrogen Atom (Section 27.4)

It is experimentally observed that a hydrogen atom can emit and absorb light only at certain wavelengths, not at all wavelengths. The small number of visible wavelengths the hydrogen atom can emit and absorb (called the **Balmer series**) are given by an empirical equation $\frac{1}{\lambda} = R\left(\frac{1}{2^2} - \frac{1}{n^2}\right)$ for $n = 3, 4, 5,$ and 6, where $R = 1.097 \times 10^{-2}$ nm^{-1} is called the *Rydberg constant*. The **Bohr theory of the hydrogen atom** successfully explained the **emission spectrum** (a series of bright lines) and **absorption spectrum** (a series of dark lines superimposed on a continuous spectrum) of the hydrogen atom. In Bohr's theory, he assumed that

- the hydrogen electron orbits the nuclear proton in a circular orbit (analogous to planets orbiting the Sun);
- the angular momentum of the electron is quantized in integral multiples of Planck's constant, h;
- the electron does not radiate energy when it is in certain discrete circular orbits;
- the electron radiates or absorbs energy only when it makes a transition to another orbit.

From these assumptions, Bohr showed that the electron can have only certain size orbits with certain energies. The energies and the radii of the orbits are given by $E_n = -\frac{13.6\text{ eV}}{n^2}$ and $r_n = 0.0529n^2$ nm for $n = 1, 2, 3, 4, \ldots$, where n is an example of a *quantum number*, specifically the **principal quantum number**. The $n = 1$ orbit is known as the **ground state**, and orbits with $n > 1$ are called the **excited states**. The energy of the electron in any state is E_n, and the energy needed to completely free the electron from the atom in that state is $-E_n$, which is called the **binding energy**.

An electron generally does not remain in an excited state for long. It *decays*, or makes a transition to a lower energy level, in a short time. The time an electron spends in an excited state is called the **lifetime** of the excited state. If an electron makes a downward transition from the n_i state to the n_f state, a photon is released, and its energy is equal to the energy difference between the final and initial states, $\Delta E = E_f - E_i = 13.6\left(\dfrac{1}{n_f^2} - \dfrac{1}{n_i^2}\right)$ eV.

The wavelength of the photon is then $\lambda = \dfrac{hc}{\Delta E} = \dfrac{1.24 \times 10^3 \text{ eV·nm}}{\Delta E}$. For transitions with $n_f = 1$, the spectrum series is called the *Lyman series* (all ultraviolet), for $n_f = 2$, the *Balmer series* (visible if $n_i = 3, 4, 5$, and 6), for $n_f = 3$, the *Paschen series* (infrared), and so on.

Example 27.6   What are the orbital radius and total energy of an electron for a hydrogen atom in (a) the ground state and (b) the second excited state?

Solution: Given: (a) $n = 1$ (ground state)   (b) $n = 3$ (second excited state).   Find: r and E.

Ground state corresponds to $n = 1$, and $n = 2$ and 3 are for the first and second excited states, respectively.

(a) From $r_n = 0.0529n$ nm, $r_1 = 0.0529(1)$ nm $= 0.0529$ nm.

From $E_n = -\dfrac{13.6 \text{ eV}}{n^2}$, $E_1 = -\dfrac{13.6 \text{ eV}}{1^2} = -13.6$ eV.

(b) $r_3 = 0.0529(3)$ nm $= 0.159$ nm, and $E_3 = -\dfrac{13.6 \text{ eV}}{3^2} = -1.51$ eV.

Integrated Example 27.7

Two electrons of a hydrogen atom in the fourth excited state make transitions to the first excited state and to the ground state, respectively. (a) The photon wavelength for the transition to the first excited state is (1) longer than, (2) the same as, or (3) shorter than the photon wavelength for the transition to the ground state. Explain. (b) What are the energies and wavelengths of the emitted photons?

(a) Conceptual Reasoning:

Because the ground state corresponds to $n = 1$, the $n = 2$ and 5 are for the first and fourth excited states. The energy difference between the fourth excited state ($n = 5$) and the ground state ($n = 1$) is greater than the energy difference between the fourth excited state ($n = 5$) and the first excited state ($n = 2$). Since energy is directly proportional to frequency, it is inversely proportional to wavelength. That is, the greater the energy difference, the shorter the wavelength. Therefore, the photon wavelength for the transition to the first excited state is (1) longer than the photon wavelength for the transition to the ground state.

(b) Quantitative Reasoning and Solution:

Given: $n_i = 5$, $n_f = 1$ and 2. Find: ΔE and λ.

Since $E_n = -\dfrac{13.6 \text{ eV}}{n^2}$, $\Delta E = E_i - E_f = 13.6 \left(\dfrac{1}{n_f^2} - \dfrac{1}{n_i^2} \right)$ eV.

Also because $\Delta E = \dfrac{1.24 \times 10^3 \text{ eV·nm}}{\lambda}$, $\lambda = \dfrac{1.24 \times 10^3 \text{ eV·nm}}{\Delta E}$.

Fourth excited state ($n = 5$) to ground state ($n = 1$):

$\Delta E_{51} = 13.6 \left(\dfrac{1}{1^2} - \dfrac{1}{5^2} \right)$ eV $= 13.06$ eV and $\lambda_{51} = \dfrac{1.24 \times 10^3 \text{ eV·nm}}{13.06 \text{ eV}} = 94.9$ nm.

Fourth excited state ($n = 5$) to first excited state ($n = 2$):

$\Delta E_{52} = 13.6 \left(\dfrac{1}{2^2} - \dfrac{1}{5^2} \right)$ eV $= 2.86$ eV and $\lambda_{52} = \dfrac{1.24 \times 10^3 \text{ eV·nm}}{2.86 \text{ eV}} = 434$ nm.

As expected, the photon wavelength for the transition to the first excited state is longer than the photon wavelength for the transition to the ground state. The emitted photon associated to the transition to the ground state is an ultraviolet photon and the one to the first excited state is in the blue/violet region of the visible spectrum.

---

5. A Quantum Success: The Laser (Section 27.5)

The quantum theory of atomic structure gave rise to the development of the **laser**, an acronym for *l*ight *a*mplification by *s*timulated *e*mission of *r*adiation. **Stimulated emission** is the process in which a photon with an energy equal to an allowed transition strikes an atom in an excited state, stimulating the atom to make a transition and emit a photon of the same energy, frequency, traveling direction, and phase as the incoming photon. In this process, one photon goes in and two come out (the incoming one and the one emitted by the atom), resulting in the amplification of that light. Another basic requirement for laser action is **population inversion**, that is, more electrons must occupy a **metastable state**, a state in which an excited electron remains for a relatively longer time, than an excited state of lower energy.

Phosphorescent materials are examples of substances made up of atoms with metastable states. Normally, there are more electrons at lower energy levels, and these can be elevated to metastable states by various "pumping" or energy input processes. Laser beams are very intense, highly directional, coherent (same phase), and monochromatic (same frequency). Depending on the lasing medium, lasers can be classified as gas lasers, solid lasers, semiconductor lasers, dye lasers, X-ray lasers, free-electron lasers, and so forth. Lasers have a wide variety of applications, ranging from compact disk players to welding and eye surgery.

An interesting application of laser light is the production of three-dimensional images in a process called **holography**. The key to holography is the coherent property of laser light, which gives the light waves a definite spatial relationship to one another. The process does not use lenses as ordinary image-forming processes do, yet it recreates the original scene in three dimensions. Holography has many potential applications in medicine, sciences, and even in our everyday life (holographic three-dimensional television).

III.   Mathematical Summary

Wien's Displacement Law	$\lambda_{max} T = 2.90 \times 10^{-3}$ m·K (27.1)	Relates the absolute temperature and the wavelength of maximum intensity.
Planck's Hypothesis	$E_n = n(hf)$, for $n = 1, 2, 3, \ldots$ $h = 6.63 \times 10^{-34}$ J·s (27.2)	States that the energy of the atoms was quantized in multiples of their vibrational frequency.
Photon Energy	$E = hf$ (27.4)	Defines the energy of a photon in terms of frequency.
Compton Scattering Equation (from free electrons)	$\Delta\lambda = \lambda - \lambda_o = \lambda_C (1 - \cos\theta)$ (27.9) $\lambda_C = h/(m_e c) = 2.43 \times 10^{-3}$ nm	Computes the wavelength shift (increase) of light scattered by a free electron or other charged particles.
Bohr Theory Orbit Radius	$R_n = 0.0529n^2$ nm For $n = 1,2,3,\ldots$ (27.16)	Calculates the orbit radius of an electron in a hydrogen atom.
Bohr Theory Electron Energy	$E_n = \dfrac{-13.6}{n^2}$ eV $n = 1, 2, 3, \ldots$ (27.17)	Calculates the total energy of an electron in a hydrogen atom.
Bohr Theory Photon Wavelength	$\lambda = \dfrac{hc}{\Delta E} = \dfrac{1.24 \times 10^3}{\Delta E \text{ (in eV)}}$ nm (27.20)	Calculates the wavelength of a photon emitted when an electron makes a transition.

IV.   Solutions of Selected Exercises and Paired Exercises

8.   From $\lambda_{max} T = 2.90 \times 10^{-3}$ m·K, we have $T = \dfrac{2.90 \times 10^{-3} \text{ m·K}}{\lambda_{max}} = \dfrac{2.90 \times 10^{-3} \text{ m·K}}{700 \times 10^{-9} \text{ m}} = \boxed{4.14 \times 10^3 \text{ K}}$.

12.   (a) The frequency of the most intense spectral component will $\boxed{\text{(1) increase, but not double}}$ if the temperature is increased from 200°C to 400°C. The temperature used should be in kelvin, and 200°C = 473 K and 400°C = 673 K. 673 K is higher than 473 K but not double.

(b) From $\lambda_{max} T = 2.90 \times 10^{-3}$ m·K, we have $\lambda_{max} = \dfrac{2.90 \times 10^{-3} \text{ m·K}}{T}$.

$$\Delta f = f_2 - f_1 = \frac{c}{\lambda_1} - \frac{c}{\lambda_2} = \frac{c\, T_2}{2.90 \times 10^{-3} \text{ m·K}} - \frac{c\, T_1}{2.90 \times 10^{-3} \text{ m·K}} = \frac{c}{2.90 \times 10^{-3} \text{ m·K}} \Delta T$$

$$= \frac{3.00 \times 10^8 \text{ m/s}}{2.90 \times 10^{-3} \text{ m·K}} \times (200 \text{ K}) = \boxed{2.07 \times 10^{13} \text{ Hz}}.$$

20. X-ray has a shorter wavelength than red light so it has a higher frequency. Therefore the X-ray photon has much more energy than the red-light photon ($E = hf$).

24. (a) A quantum of violet light has $\boxed{(1) \text{ more}}$ energy because the energy is proportional to frequency, $E = hf$, and frequency is inversely proportional to wavelength. So the shorter the wavelength, the higher the frequency and the higher the energy.

(b) From $E = hf$, we have $\dfrac{E_v}{E_r} = \dfrac{f_v}{f_r} = \dfrac{\lambda_r}{\lambda_v} = \dfrac{700 \text{ nm}}{400 \text{ nm}} = \boxed{1.75}$.

26. (a) Because $\phi_o = hf_o = \dfrac{hc}{\lambda_o}$, the threshold wavelength for metal A is $\boxed{(1) \text{ shorter than}}$. The greater the work function, the higher the threshold (cut-off) frequency and the shorter the threshold wavelength.

(b) Since $\phi_o = hf_o = \dfrac{hc}{\lambda_o}$, $\dfrac{\lambda_{oA}}{\lambda_{oB}} = \dfrac{\phi_{oB}}{\phi_{oA}} = \dfrac{1}{2}$. So $\lambda_{oA} = \dfrac{1}{2} \lambda_{oB} = \dfrac{1}{2} (620 \text{ nm}) = \boxed{310 \text{ nm}}$.

30. $hf = eV_o + \phi_o = 2.50 \text{ eV} + 2.40 \text{ eV} = 4.90 \text{ eV} = 7.84 \times 10^{-19}$ J.

So $f = \dfrac{7.84 \times 10^{-19} \text{ J}}{6.63 \times 10^{-34} \text{ J·s}} = 1.183 \times 10^{15}$ Hz.

Therefore $\lambda = \dfrac{c}{f} = \dfrac{3.00 \times 10^8 \text{ m/s}}{1.183 \times 10^{15} \text{ Hz}} = 2.54 \times 10^{-7} \text{ m} = \boxed{254 \text{ nm}}$.

34. (a) $eV_o = K_{max} = hf - \phi_o = \dfrac{hc}{\lambda} - \phi_o = \dfrac{1.24 \times 10^3 \text{ eV·nm}}{300 \text{ nm}} - 3.5 \text{ eV} = 0.63 \text{ eV}.$

So $V_o = \boxed{0.63 \text{ V}}$.

(b) $f_o = \dfrac{\phi_o}{h} = \dfrac{(3.5 \text{ eV})(1.6 \times 10^{-19} \text{ J/eV})}{6.63 \times 10^{-34} \text{ J·s}} = \boxed{8.4 \times 10^{14} \text{ Hz}}$.

43. This is to conserve total linear momentum, which is a vector quantity. Since the initial momentum has no component in the y-direction, the electron must have a y-component in the opposite side.

46. $\Delta\lambda = \lambda_C(1 - \cos\theta) = (0.00243\text{ nm})(1 - \cos 30°) = \boxed{3.26 \times 10^{-4}\text{ nm}}$.

52. (a) The Compton wavelength for a proton is $\boxed{(3)\text{ shorter}}$ because it depends inversely on mass, $\lambda_C = \dfrac{h}{mc}$, compared with the Compton wavelength for an electron.

(b) $\lambda_C = \dfrac{h}{m_p c^2} = \dfrac{6.63 \times 10^{-34}\text{ J·s}}{(1.67 \times 10^{-27}\text{ kg})(3.00 \times 10^8\text{ m/s})} = \boxed{1.32 \times 10^{-15}\text{ m}}$.

(c) The maximum wavelength shift is equal to $2\lambda_C$ when $\theta = 180°$, from $\Delta\lambda = \lambda_C(1 - \cos\theta)$.

$$\frac{(\Delta\lambda_e)_{max}}{(\Delta\lambda_p)_{max}} = \frac{2\lambda_{Ce}}{2\lambda_{Cp}} = \frac{2.43 \times 10^{-12}\text{ m}}{1.32 \times 10^{-15}\text{ m}} = \boxed{1.84 \times 10^3}.$$

60. (a) $\Delta E = (-13.6\text{ eV})\left(\dfrac{1}{n_f^2} - \dfrac{1}{n_i^2}\right) = (-13.6\text{ eV})\left(\dfrac{1}{\infty} - \dfrac{1}{2^2}\right) = \boxed{3.40\text{ eV}}$.

(b) $\Delta E = (-13.6\text{ eV})\left(\dfrac{1}{\infty} - \dfrac{1}{3^2}\right) = \boxed{1.51\text{ eV}}$.

62. (a) Since $r_n = 0.0529n^2$ nm, $r_3 = (0.0529)(2)^2$ nm $= \boxed{0.212\text{ nm}}$.

(b) $r_6 = 0.0529(4)^2 = \boxed{0.846\text{ nm}}$.

(c) $r_{10} = 0.0529(5)^2$ nm $= \boxed{1.32\text{ nm}}$.

66. (a) $\Delta E = (-13.6\text{ eV})\left(\dfrac{1}{n_f^2} - \dfrac{1}{n_i^2}\right)$. $\Delta E_{52} = (13.6\text{ eV})\left(\dfrac{1}{2^2} - \dfrac{1}{5^2}\right) = 2.856\text{ eV}$.

$\lambda_{52} = \dfrac{1.24 \times 10^3}{\Delta E\text{ (in eV)}}$ nm $= \dfrac{1.24 \times 10^3}{2.856}$ nm $= \boxed{434\text{ nm}}$. $\Delta E_{21} = (13.6\text{ eV})\left(\dfrac{1}{1^2} - \dfrac{1}{2^2}\right) = 10.2\text{ eV}$.

$\lambda_{21} = \dfrac{1.24 \times 10^3}{10.2}$ nm $= \boxed{122\text{ nm}}$.

(b) $\boxed{\text{Yes, the transition from } n = 5 \text{ to } n = 2 \text{ is in the visible region}}$ (violet).

72. The kinetic energy is $K = \dfrac{ke^2}{2r}$ and the potential energy is $U = -\dfrac{ke^2}{r}$.

Thus, the total energy is

$$E = U + K = \frac{ke^2}{2r} - \frac{ke^2}{r} = -\frac{ke^2}{2r} = -\frac{(9.0 \times 10^9\text{ N·m}^2/\text{C}^2)(1.6 \times 10^{-19}\text{ C})^2}{2(0.0529 \times 10^{-9}\text{ m})}$$

$$= -2.18 \times 10^{-18}\text{ J} = -13.6\text{ eV}.$$

79. In a spontaneous emission, electrons jump from a higher-energy state to a lower-energy state without any external stimulation, and a photon is released in the process.

Stimulated emission is an induced emission. The electron in the higher-energy orbit can jump to a lower-energy orbit when a photon of energy equal to the difference of the energy between the two orbits is introduced. Once atoms are prepared with enough electrons in the higher energy state, stimulating photons trigger them to jump down to the lower-energy state. The emitted photons trigger the rest of the electrons and eventually all the electrons will be in the lower energy state.

82. (a) The answer is $\boxed{\text{(1) one}}$.

(b) It is from $n = \boxed{2 \text{ to } 3}$.

(c) $\Delta E = (-13.6 \text{ eV}) \left(\dfrac{1}{n_f^2} - \dfrac{1}{n_i^2} \right) = (-13.6 \text{ eV}) \left(\dfrac{1}{3^2} - \dfrac{1}{2^2} \right) = \boxed{1.89 \text{ eV}}$.

$\lambda = \dfrac{1.24 \times 10^3 \text{ eV·nm}}{\Delta E} = \dfrac{1.24 \times 10^3 \text{ eV·nm}}{1.89 \text{ eV}} = \boxed{656 \text{ nm}}$ (red).

V. Practice Quiz

1. Which of the following colors is associated with a blackbody of the lowest temperature?

(a) red (b) yellow (c) orange (d) green (e) blue

2. If the wavelength of a photon is doubled, by what factor does the energy change?

(a) 4 (b) 2 (c) 1 (d) 1/2 (e) 1/4

3. A hypothetical atom has two excited states in addition to its ground state. How many different spectral lines are possible?

(a) 2 (b) 3 (c) 4 (d) 5 (e) 6

4. The kinetic energy of a photoelectron depends on which one of the following?

(a) intensity of light (b) duration of illumination (c) stopping potential

(d) wavelength of light (e) angle of illumination

5. If the scattering angle in Compton scattering from an electron is 45°, what is the wavelength shift?

(a) 7.11×10^{-4} m (b) 1.72×10^{-3} m (c) 2.43×10^{-3} m (d) 3.44×10^{-3} m (e) 6.08×10^{-3} m

6.   What is the frequency of the most intense radiation from an object with temperature 100°C?

          (a) 7.8×10^{-6} Hz   (b) 2.9×10^{-5} Hz   (c) 3.9×10^{13} Hz   (d) 1.0×10^{13} Hz   (e) 1.0×10^{11} Hz

7.   A monochromatic light beam is incident on the surface of a metal with work function 2.50 eV. If a stopping potential of 1.0 V is required to make the photocurrent zero, what is the wavelength of the light?

          (a) 1.42×10^{3} nm   (b) 744 nm   (c) 497 nm   (d) 423 nm   (e) 354 nm

8.   The binding energy of the hydrogen atom in its ground state is 13.6 eV. What is the energy of the atom when it is in the $n = 5$ state?

          (a) 2.72 eV   (b) −2.72 eV   (c) 0.544 eV   (d) −0.544 eV   (e) −13.6 eV

9.   A hydrogen atom in its ground state absorbs a photon of energy 12.09 eV. To which state will the electron make a transition?

          (a) $n = 1$ (b) $n = 2$ (c) $n = 3$ (d) $n = 4$ (e) $n = 5$

10.   The wavelength of a He-Ne laser is 632.8 nm. What is the energy difference, in eV, between the two energy states involved in producing this laser action?

          (a) 0.509 eV   (b) 1.96 eV   (c) 3.14 eV   (d) 4.74 eV   (e) 13.6 eV

11.   At what scattering angle will the photon in a Compton scattering experiment undergo the greatest change in its wavelength?

          (a) 0°   (b) 45°   (c) 90°   (d) 135°   (e) 180°

12.   An FM radio station broadcasts at a frequency of 98.9 MHz and radiates 750 kW of power. How many photons are radiated by the station's antenna each second?

          (a) 1.14×10^{26} (b) 6.56×10^{26} (c) 1.14×10^{28} (d) 6.56×10^{28} (e) 1.14×10^{31}

Answers to Practice Quiz:

1.a 2.d 3.b 4.d 5.a 6.c 7.e 8.d 9.c 10.b 11.e 12.e

CHAPTER 28

Quantum Mechanics and Atomic Physics

I.   Chapter Objectives

Upon completion of this chapter, you should be able to:

1.   explain de Broglie's hypothesis, calculate the wavelength of a matter wave, and specify under what circumstances the wave nature of matter will be observable.

2.   understand qualitatively the reasoning that underlies the Schrödinger wave equation, and the equation's use in finding particle wave functions.

3.   understand the structure of the periodic table in terms of quantum mechanical electron orbits, and the Pauli exclusion principle.

4.   understand the inherent quantum-mechanical limits on the accuracy of physical observations.

5.   understand the relationship between particles and antiparticles, and the energy requirements for pair production.

II.   Chapter Summary and Discussion

1.   Matter Waves: The de Broglie Hypothesis (Section 28.1)

The **de Broglie hypothesis** associates wavelength with moving material particles by reverse analogy with the assignment of particle nature to a wave of light. From relativity, the magnitude of the momentum of a photon (which has no mass) is $p = \dfrac{E}{c} = \dfrac{hf}{c} = \dfrac{h}{\lambda}$. The de Broglie hypothesis states that whenever a particle has momentum of magnitude p there is a wave associated with it. In analogy with a light photon, that wave should have a wavelength of $\lambda = \dfrac{h}{p} = \dfrac{h}{mv}$. These waves associated with moving particles are called **matter waves** or, more commonly, **de Broglie waves**. Matter waves of particles have been experimentally verified by the *Davisson-Germer* experiment (electron diffraction) and the *Thomson* experiment (also electron diffraction).

Note: For convenience, the wavelength of the matter wave of an electron accelerated from rest through a potential V (in volts) can be calculated from $\lambda = \sqrt{\dfrac{1.50}{V}}$ nm.

Example 28.1   What is the wavelength of the matter wave associated with

(a) a tennis ball of mass 0.057 kg moving with a speed of 25 m/s?

(b) an electron moving with a speed of 2.5×10^7 m/s?

Solution: Given: (a) $m = 0.057$ kg, $v = 25$ m/s.   (b) $m = 9.11 \times 10^{-31}$ kg, $v = 2.5 \times 10^7$ m/s.

Find: (a) and (b) λ.

(a) $\lambda = \dfrac{h}{p} = \dfrac{h}{mv} = \dfrac{6.63 \times 10^{-34} \text{ J·s}}{(0.057 \text{ kg})(25 \text{ m/s})} = 4.7 \times 10^{-34}$ m.

(b) $\lambda = \dfrac{6.63 \times 10^{-34} \text{ J·s}}{(9.11 \times 10^{-31} \text{ kg})(2.5 \times 10^7 \text{ m/s})} = 2.9 \times 10^{-11}$ m $= 290$ nm.

The wavelength of the tennis ball is much shorter than that of the electron. That is why matter waves are important for small particles like electrons, since their wavelengths are comparable to the sizes of the objects with which they interact. It is easier to observe interference and diffraction for electrons than for the tennis ball and all other everyday objects.

Example 28.2   An electron is accelerated by a potential difference of 120 V.

(a) What is the wavelength of the matter wave associated with the electron?

(b) What is the momentum of the electron?

(c) What is the kinetic energy of the electron?

Solution: Given: $m = 9.11 \times 10^{-31}$ kg, $V = 120$ V.

Find: (a) λ (b) p (c) K.

(a) $\lambda = \sqrt{\dfrac{1.50}{V}}$ nm $= \sqrt{\dfrac{1.50}{120}}$ nm $= 0.112$ nm.

(b) From $\lambda = \dfrac{h}{p}$, we have $p = \dfrac{h}{\lambda} = \dfrac{6.63 \times 10^{-34} \text{ J·s}}{0.112 \times 10^{-9} \text{ m}} = 5.92 \times 10^{-24}$ kg·m/s.

(c) $K = \frac{1}{2}mv^2 = \dfrac{p^2}{2m} = \dfrac{(5.92 \times 10^{-24} \text{ kg·m/s})^2}{2(9.11 \times 10^{-31} \text{ kg})} = 1.92 \times 10^{-17}$ J.

Here we used $p = mv$.

2. The Schrödinger Wave Equation (Section 28.2)

The **Schrödinger wave equation** is an equation that enables us to calculate matter waves for particles in various systems like atoms. The function satisfying the wave equation is called the **wave function**, ψ, which describes the wave as a function of time and space. The general form of the wave equation can be written as $(K + U)\psi = E\psi$, where K, U, and E are the kinetic, potential, and total energy of the particle, respectively. The square of the wave function (ψ^2) is called the **probability density**, which represents the relative probability of finding a particle in space and time. The interpretation of ψ^2 as the probability of finding a particle at a particular place altered the idea that particles are found in certain definite locations. Now we can say that only a particle has a certain probability of being at some location and cannot predict exactly where it is or will be. In some instances, a particle has a finite probability of being even in a classically forbidden region. For example, the classical locations of particles such as the electron orbits in a hydrogen atom are only the *most probable* locations of the particles.

The solutions to the Schrödinger wave equation lead to many experimentally proven quantum effects (classically forbidden) such as *tunneling* through a *potential energy barrier*. A classical analogy to quantum tunneling would be that a person could run though a concrete wall if he/she tried enough "hits." It takes a lot of energy, say E, to go through the concrete wall, and the kinetic energy of a human being is certainly smaller than E. Therefore, classically the person cannot run through the concrete wall because he/she does not have enough energy to overcome the energy barrier of the wall; however, according to quantum mechanics, there is a small probability that the person might appear on the other side of the wall. It is possible for the impossible (in classical sense) to happen if it is tried enough times. The electron tunneling microscope is an application of this quantum effect.

3. Atomic Quantum Numbers and the Periodic Table (Section 28.3)

When the Schrödinger wave equation is solved for the hydrogen atom, the solution gives three quantum numbers: the principal quantum number n, the orbital quantum number ℓ, and the magnetic quantum number m_ℓ. The spin quantum number m_s was added to agree with experimental observations. The **principal quantum number** is the sole determiner of the energy levels of a hydrogen atom as predicted by the Bohr theory (in the absence of external magnetic field). The **orbital quantum number** determines the allowed value of angular momentum for an electron orbit. The **magnetic quantum number** determines the orientation of the plane of the electron orbit with respect to a given axis. The **spin quantum number** is a purely quantum-mechanical concept, that, in a classical analogy, indicates whether the spin angular momentum of an electron is up or down relative to a given axis.

Numerically, the principal quantum number can be any positive integer, $n = 1, 2, 3, \ldots, \infty$. For a given n, the orbital quantum number can be $\ell = 0, 1, 2, 3, \ldots, (n-1)$, or n different values. For example, if $n = 2$, $\ell = 0$ and 1 (two different values); if $n = 5$, $\ell = 0, 1, 2, 3$, and 4 (five different values). For a given ℓ, the magnetic quantum number can be $m_\ell = 0, \pm 1, \pm 2, \pm 3, \ldots, \pm \ell$, or $2\ell + 1$ different values. The spin quantum number can be either $+\frac{1}{2}$ (spin up) or $-\frac{1}{2}$ (spin down). Each set of unique quantum numbers (n, ℓ, m_ℓ, m_s) defines a quantum state.

It is common to refer to the total of all the states in a given n as forming a **shell**, and to ℓ levels of that shell as **subshells**. That is, atomic electrons with the same n value are said to be in the same shell. Electrons with the same ℓ value are said to be in the same subshell. The subshells are designated by the letters $s, p, d, f, g, \ldots$ for $\ell = 0, 1, 2, 3, 4, \ldots$, respectively.

The **Pauli exclusion principle** states that no two electrons in a multielectron atom can have the same set of quantum numbers (n, ℓ, m_ℓ, m_s). That is, no two electrons can be in the same quantum state.

Example 28.3   Write down the possible sets of quantum numbers for each individual quantum state in the $n = 4$ shell.

Solution:

For each n, $\ell = 0, 1, 2, 3, \ldots, n-1$. For each ℓ, $m_\ell = 0, \pm 1, \pm 2, \pm 3, \ldots, \pm \ell$. m_s can be $\pm \frac{1}{2}$.

For $n = 4$, $\ell = 0, 1, 2$, and 3.

For $\ell = 0$, $m_\ell = 0$, and $m_s = +\frac{1}{2}$ or $-\frac{1}{2}$.

Therefore, there are two states, $(4, 0, 0, +\frac{1}{2})$ and $(4, 0, 0, -\frac{1}{2})$, for $\ell = 0$.

For $\ell = 1$, $m_\ell = 0$, and ± 1.

For $m_\ell = 0$, $m_s = +\frac{1}{2}$ or $-\frac{1}{2}$, so there are two states, $(4, 1, 0, +\frac{1}{2})$ and $(4, 1, 0, -\frac{1}{2})$.

For $m_\ell = +1$, $m_s = +\frac{1}{2}$ or $-\frac{1}{2}$, so there are two states, $(4, 1, 1, +\frac{1}{2})$ and $(4, 1, 1, -\frac{1}{2})$.

For $m_\ell = -1$, $m_s = +\frac{1}{2}$ or $-\frac{1}{2}$, so there are two states, $(4, 1, -1, +\frac{1}{2})$ and $(4, 1, -1, -\frac{1}{2})$.

Therefore, there are six states for $\ell = 1$.

For $\ell = 2$, $m_\ell = 0$, ± 1, ± 2.

For $m_\ell = 0$, $m_s = +\frac{1}{2}$ or $-\frac{1}{2}$, so there are two states, $(4, 2, 0, +\frac{1}{2})$ and $(4, 2, 0, -\frac{1}{2})$.

For $m_\ell = +1$, $m_s = +\frac{1}{2}$ or $-\frac{1}{2}$, so there are two states, $(4, 2, 1, +\frac{1}{2})$ and $(4, 2, 1, -\frac{1}{2})$.

For $m_\ell = -1$, $m_s = +\frac{1}{2}$ or $-\frac{1}{2}$, so there are two states, $(4, 2, -1, +\frac{1}{2})$ and $(4, 2, -1, -\frac{1}{2})$.

For $m_\ell = +2$, $m_s = +\frac{1}{2}$ or $-\frac{1}{2}$, so there are two states, $(4, 2, 2, +\frac{1}{2})$ and $(4, 2, 2, -\frac{1}{2})$.

For $m_\ell = -2$, $m_s = +\frac{1}{2}$ or $-\frac{1}{2}$, so there are two states, $(4, 2, -2, +\frac{1}{2})$ and $(4, 2, -2, -\frac{1}{2})$.

Therefore, there are ten states for $\ell = 2$.

For $\ell = 3$, $m_\ell = 0$, ± 1, ± 2, ± 3.

For $m_\ell = 0$, $m_s = +\frac{1}{2}$ or $-\frac{1}{2}$, so there are two states, $(4, 3, 0, +\frac{1}{2})$ and $(4, 3, 0, -\frac{1}{2})$.

For $m_\ell = +1$, $m_s = +\frac{1}{2}$ or $-\frac{1}{2}$, so there are two states, $(4, 3, 1, +\frac{1}{2})$ and $(4, 3, 1, -\frac{1}{2})$.

For $m_\ell = -1$, $m_s = +\frac{1}{2}$ or $-\frac{1}{2}$, so there are two states, $(4, 3, -1, +\frac{1}{2})$ and $(4, 3, -1, -\frac{1}{2})$.

For $m_\ell = +2$, $m_s = +\frac{1}{2}$ or $-\frac{1}{2}$, so there are two states, $(4, 3, 2, +\frac{1}{2})$ and $(4, 3, 2, -\frac{1}{2})$.

For $m_\ell = -2$, $m_s = +\frac{1}{2}$ or $-\frac{1}{2}$, so there are two states, $(4, 3, -2, +\frac{1}{2})$ and $(4, 3, -2, -\frac{1}{2})$.

For $m_\ell = +3$, $m_s = +\frac{1}{2}$ or $-\frac{1}{2}$, so there are two states, $(4, 3, 3, +\frac{1}{2})$ and $(4, 3, 3, -\frac{1}{2})$.

For $m_\ell = -3$, $m_s = +\frac{1}{2}$ or $-\frac{1}{2}$, so there are two states, $(4, 3, -3, +\frac{1}{2})$ and $(4, 3, -3, -\frac{1}{2})$.

Therefore, there are 14 states for $\ell = 3$.

Hence, there are a total of $2 + 6 + 10 + 14 = 32$ states.

Generally, there are $2n^2$ states for a given value of n. Here $2n^2 = 2(4)^2 = 32$.

Example 28.4   How many possible sets of quantum numbers or electron states are there in the 5*f* subshell? What is the maximum number of electrons that can occupy this subshell?

Solution:

For the 5*f* subshell, $n = 5$ and $\ell = 3$.

For $\ell = 3$, m_ℓ can be 0, ± 1, ± 2, ± 3, or seven different values, $(2\ell + 1) = (2 \times 3 + 1) = 7$ values.

For each m_ℓ, m_s can be either $+\frac{1}{2}$ or $-\frac{1}{2}$, so there are 2 values.

Therefore, there are $2 \times 7 = 14$ sets of quantum numbers or electron states.

According to the Pauli exclusion principle, only one electron can occupy an electron state, so the maximum number of electrons that can occupy the 5f subshell is 14.

In a shorthand notation called **electron configuration**, we write the quantum states in increasing order of energy, and designate the numbers of electrons in each level with a subscript. For example, $4p^5$ means that the $4p$ subshell contains five electrons. The electron configuration for the ground state of the element neon (Ne), for example, is $1s^2 2s^2 2p^6$ (both shells $n = 1$, $n = 2$ completely full).

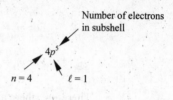

Some energy levels, such as 2s and 2p, have similar energies. We refer to such sets of energy levels that have about the same energy as an **electron period**. These electron periods are the basis of the **periodic table of elements**. In a periodic table, elements are arranged in horizontal rows, which are called **periods**, in order of increasing atomic mass. Elements of similar chemical properties are arranged in vertical columns called **groups**, or families of elements with similar properties.

4. The Heisenberg Uncertainty Principle (Section 28.4)

The **Heisenberg uncertainty principle** as applied to position (x) and momentum (magnitude p) may be stated as follows: it is impossible to know simultaneously an object's exact position and momentum. This principle overthrows the *deterministic* view of nature in classical physics. The minimum uncertainty of the product of position and momentum is $(\Delta p)(\Delta x) \geq \dfrac{h}{2\pi}$, which is expressed alternatively in energy and time as $(\Delta E)(\Delta t) \geq \dfrac{h}{2\pi}$.

Note: The value $h/(2\pi)$ is the minimum uncertainty of the product of position and momentum, that is, *at the very best*, the uncertainty of the product is $h/(2\pi)$, that is, it can never be zero.

Integrated Example 28.5

Two measurements of an electron's speed are both 1.0×10^6 m/s. Measurement A has an uncertainty of $\pm 10\%$ and measurement B has an uncertainty of $\pm 5.0\%$. (a) The uncertainty associated with measurement A is (1) greater than, (2) the same as, or (3) less than that associated with measurement B. Explain. (b) What are the minimum uncertainties in its position from the two measurements?

(a) Conceptual Reasoning:

From the uncertainty principle, $(\Delta p)(\Delta x) \geq \dfrac{h}{2\pi}$, we see that the product of the uncertainty in position and the uncertainty in momentum is a constant. Since $p = mv$, the uncertainty in speed is thus inversely proportional to the uncertainty in position. Measurement A has a greater uncertainty in speed so its uncertainty in position should be (3) less than that associated with measurement B.

(b) Quantitative Reasoning and Solution:

Given: $m = 9.11 \times 10^{-31}$ kg, $v = 1.0 \times 10^6$ m/s,

% uncertainty in $v_A = \pm 10\%$,   % uncertainty in $v_B = \pm 5.0\%$.

Find: Δx for both measurements A and B.

Measurement A: $v_{max} = v + 0.10v = 1.0 \times 10^6$ m/s $+ (0.10)(1.0 \times 10^6$ m/s$) = 1.1 \times 10^6$ m/s;

$v_{min} = v - 0.10v = 1.0 \times 10^6$ m/s $- (0.10)(1.0 \times 10^6$ m/s$) = 0.90 \times 10^6$ m/s.

Therefore, the uncertainty in momentum (the range in momentum values) is

$\Delta p_A = mv_{max} - mv_{min} = (9.11 \times 10^{-31}$ kg$)(1.1 \times 10^6$ m/s $- 0.90 \times 10^6$ m/s$) = 1.8 \times 10^{-25}$ kg·m/s.

Thus, $\Delta x_A \geq \dfrac{h}{2\pi (\Delta p_A)} = \dfrac{6.63 \times 10^{-34} \text{ J·s}}{2\pi (1.8 \times 10^{-25} \text{ kg·m/s})} = 5.9 \times 10^{-10}$ m $= 0.59$ nm $\approx$ atomic dimensions.

Measurement B: $v_{max} = v + 0.050v = 1.0 \times 10^6$ m/s $+ (0.050)(1.0 \times 10^6$ m/s$) = 1.05 \times 10^6$ m/s;

$v_{min} = v - 0.050v = 1.0 \times 10^6$ m/s $- (0.050)(1.0 \times 10^6$ m/s$) = 0.95 \times 10^6$ m/s.

$\Delta p = (9.11 \times 10^{-31}$ kg$)(1.05 \times 10^6$ m/s $- 0.95 \times 10^6$ m/s$) = 9.1 \times 10^{-26}$ kg·m/s.

Thus, $\Delta x_B \geq \dfrac{6.63 \times 10^{-34} \text{ J·s}}{2\pi (9.1 \times 10^{-26} \text{ kg·m/s})} = 1.2 \times 10^{-9}$ m $= 1.2$ nm.

As expected, measurement A has a smaller uncertainty in position than measurement B.

Example 28.6 The energy of an electron in an atomic state has an uncertainty of about ± 0.0500 eV. What is the lifetime (the time the electron remains in that level before making a transition to another level) of that level?

Solution: Given: $\Delta E = 2 \times 0.0500$ eV $= 0.100$ eV $= 1.60 \times 10^{-20}$ J.

Find: Δt.

From the uncertainty principle, $(\Delta E)(\Delta t) \geq \dfrac{h}{2\pi}$, we have

$\Delta t \geq \dfrac{h}{2\pi (\Delta E)} = \dfrac{6.63 \times 10^{-34} \text{ J·s}}{2\pi (1.60 \times 10^{-20} \text{ J})} = 4.14 \times 10^{-14}$ s.

---

5. Particles and Antiparticles (Section 28.5)

A **positron** is a particle that has the same mass as an electron but possesses a *positive* electronic charge. The oppositely charged positron is said to be the **antiparticle** of the electron. All subatomic particles have antiparticles. A positron can be created only with the simultaneous creation of an electron in a process called **pair production**. In this process, the energy of a high-energy photon is converted to a positron and an electron, and afterward they have kinetic energies. By mass-energy equivalence, the *threshold energy for pair production* (the minimum energy of the proton) required is $E_{min} = hf = 2m_e c^2 = 1.022$ MeV, that is, $E \geq 1.022$ MeV. A positron and an electron can also be "destroyed" or annihilated in a process called **pair annihilation**, which is the direct conversion of mass into electromagnetic energy (photons)—the inverse of pair production so to speak. It is conceivable that antiparticles predominate in some parts of the universe. If so, the atoms of the **antimatter** in this region would consist of negatively charged nuclei (composed of antiprotons and antineutrons), surrounded by positively charged positrons (antielectrons). It would be impossible to distinguish between antimatter and ordinary matter, except that, however, if antimatter and ordinary matter should come into contact, they would annihilate each other with an explosive release of energy.

III.   Mathematical Summary

Magnitude of the Momentum of a Photon	$p = \dfrac{E}{c} = \dfrac{hf}{c} = \dfrac{h}{\lambda}$ (28.1)	Defines the magnitude of the momentum of a photon.
de Broglie Wavelength of a Moving Particle	$\lambda = \dfrac{h}{p} = \dfrac{h}{mv}$ (28.2)	Defines the deBroglie wavelength of a moving particle (matter wave) in terms of its momentum.
Heisenberg Uncertainty Principle (two forms)	$(\Delta p)(\Delta x) \geq \dfrac{h}{2\pi}$ (28.5)	Expresses the Heisenberg uncertainty principle for momentum-position and energy-time.
	$(\Delta E)(\Delta t) \geq \dfrac{h}{2\pi}$ (28.6)	

IV.   Solutions of Selected Exercises and Paired Exercises

6.   The wavelength will be   shorter, as a higher difference in potential causes a higher momentum . The de Broglie wavelength is inversely proportional to the momentum, $\lambda = \dfrac{h}{p}$.

8. (a) The electron will have $\boxed{\text{(3) a longer}}$ de Broglie wavelength due to its smaller mass. The de Broglie

wavelength is inversely proportional to the mass, $\lambda = \dfrac{h}{mv}$.

(b) $\lambda_{electron} = \dfrac{h}{mv} = \dfrac{6.63 \times 10^{-34} \text{ J·s}}{(9.11 \times 10^{-31} \text{ kg})(100 \text{ m/s})} = \boxed{7.28 \times 10^{-6} \text{ m}}$.

$\lambda_{proton} = \dfrac{6.63 \times 10^{-34} \text{ J·s}}{(1.67 \times 10^{-27} \text{ kg})(100 \text{ m/s})} = \boxed{3.97 \times 10^{-9} \text{ m}}$.

10. (a) If the potential difference increases to nine times its original value, the new de Broglie wavelength will

be $\boxed{\text{(4) one-third}}$ times that of the original. According to Eq. 28.3, $\lambda = \sqrt{\dfrac{1.50}{V}}$ nm, we can see that the

de Broglie wavelength is inversely proportional to the square root of potential difference.

(b) From Eq. 28.3: $\lambda = \sqrt{\dfrac{1.50}{V}}$ nm, so λ is proportional to $\dfrac{1}{\sqrt{V}}$.

Therefore $\dfrac{\lambda_2}{\lambda_1} = \sqrt{\dfrac{V_1}{V_2}} = \sqrt{\dfrac{250 \text{ kV}}{600 \text{ kV}}} = \boxed{0.645}$.

13. (a) Its de Broglie wavelength will $\boxed{\text{(3) decrease}}$ due to the potential difference. The proton gains speed

from the potential difference and the de Broglie wavelength is inversely proportional to speed, $\lambda = \dfrac{h}{mv}$.

(b) The initial kinetic energy of the proton is

$K_o = \frac{1}{2}mv_o^2 = \frac{1}{2}(1.67 \times 10^{-27} \text{ kg})(4.5 \times 10^4 \text{ m/s})^2 = 1.69 \times 10^{-18} \text{ J} = 10.6 \text{ eV}$.

Since the initial speed of the proton is 4.5×10^4 m/s, it had taken 10.6 V to accelerate the proton from rest
to this speed.

So $V_1 = 10.6$ V and $V_2 = 10.6$ V + 37 V = 47.6 V.

From Eq. 28.3: $\lambda = \sqrt{\dfrac{1.50}{V}}$ nm, therefore λ is proportional to $\dfrac{1}{\sqrt{V}}$.

Thus, the percentage difference is

$\dfrac{\lambda_2 - \lambda_1}{\lambda_1} = \dfrac{\lambda_2}{\lambda_1} - 1 = \sqrt{\dfrac{V_1}{V_2}} - 1 = \sqrt{\dfrac{10.6 \text{ V}}{47.6 \text{ V}}} - 1 = -0.53 = \boxed{-53\% \text{ (a decrease)}}$.

24. The following standing waves can be set up in the well:

$L = \dfrac{n\lambda_n}{2}$, $n = 1, 2, 3, \ldots$ or $\lambda_n = \dfrac{2L}{n}$.

$K_n = \dfrac{p^2}{2m} = \dfrac{(h/\lambda_n)^2}{2m} = \dfrac{h^2}{2m\lambda_n^2} = \dfrac{h^2}{2m(2L/n)^2} = n^2 \dfrac{h^2}{8mL^2}$.

29. The principle quantum number, n, gives information about the $\boxed{\text{total energy and orbit radius}}$ of an orbit in a hydrogen atom.

32. (a) $2n^2 = 2(2)^2 = \boxed{8}$ and $2(3)^2 = \boxed{18}$.

(b) For $n = 2$:

(2, 1, 1, +1/2);   (2, 1, 1, –1/2);   (2, 1, 0, +1/2);   (2, 1, 0, –1/2);

(2, 1, –1, +1/2);   (2, 1, –1, –1/2);   (2, 0, 0, +1);   (2, 0, 0, –1/2).

For $n = 3$:

(3, 2, 2, +1/2);   (3, 2, 2, –1/2);   (3, 2, 1, +1/2);   (3, 2, 1, –1/2);

(3, 2, 0, +1/2);   (3, 2, 0, –1/2);   (3, 2, –1, +1/2);   (3, 2, –1, –1/2);

(3, 2, –2, +1/2);   (3, 2, –2, –1/2);   (3, 1, 1, +1);   (3, 1, 1, –1/2);

(3, 1, –1, +1/2);   (3, 1, –1, –1/2);   (3, 1, 0, +1/2);   (3, 1, 0, –1/2);

(3, 0, 0, +1/2);   (3, 0, 0, –1/2).

36.

40. (a) If there were no electron spin, the $1s$ state would contain a maximum of $\boxed{\text{(2) one}}$ electron because one set of a unique quantum number set can only have one electron.

(b) The s state can have one electron, and the p state can have 3 electrons.

$1s^1$ is $\boxed{\text{hydrogen}}$ and $1s^1 2s^1 2p^3$ is $\boxed{\text{boron}}$.

44. Due to the uncertainty principle, the product of the uncertainty in position and the uncertainty in momentum (or mass times the uncertainty in velocity) cannot be zero. So there will always be uncertainly in both position and velocity.

48. Since $(\Delta p)(\Delta x) \geq \dfrac{h}{2\pi}$ and $p = mv$,

$$\Delta v = \frac{\Delta p}{m} \geq \frac{h}{2\pi(\Delta x)m} = \frac{6.63 \times 10^{-34} \text{ J·s}}{2\pi[(0.10 - 0.050) \times 10^{-9} \text{ m}](9.11 \times 10^{-31} \text{ kg})} = \boxed{2.3 \times 10^6 \text{ m/s}}.$$

52. From the uncertainty principle, $(\Delta E)(\Delta t) \geq \dfrac{h}{2\pi}$, we have

$$\Delta t \geq \frac{h}{2\pi(\Delta E)} = \frac{6.63 \times 10^{-34} \text{ J·s}}{2\pi(2 \times 0.0003 \text{ eV})(1.6 \times 10^{-19} \text{ J/eV})} = \boxed{1.1 \times 10^{-12} \text{ s}}.$$

58. 2 electrons and 2 positrons could work but the energy of the photon needs to be $4mc^2 \approx 2$ MeV.

 2 electrons and 1 positron will not work because charge is not conserved in the process.

60. The rest energy of the electron pair is 2×0.511 MeV $= 1.022$ MeV.

 From the conservation of energy and momentum, each photon will move in opposite direction with the

 energy. So each will carry half the total energy or $\boxed{0.511 \text{ MeV}}$.

62. (a) The pair production of a muon and an antimuon requires a photon of $\boxed{\text{(1) more}}$ energy than electron-

 positron pair because the muons have more rest mass, so more energy is required.

 (b) $E = hf \geq 2m_\mu c^2 = 207(2m_e c^2) = 207(1.022 \text{ MeV}) = \boxed{212 \text{ MeV}}$.

64. (a) There are $\boxed{\text{(3) four}}$ different spin orientations, 3/2, 1/2, –1/2, and –3/2.

 (b) You could put $\boxed{\text{four}}$ in the ground state.

V.   Practice Quiz

1. If the accuracy in measuring the velocity of a particle decreases, the accuracy in measuring its position will

 (a) increase   (b) decrease   (c) remain the same   (d) be uncertain   (e) be exact

2. What is the electron configuration for a neutral Li atom?

 (a) $1s^3$ (b) $1s^1 2s^2$ (c) $1s^2 2s^1$ (d) $1s^2 1p^1$ (e) $1s^1 1p^2$

3. An electron is accelerated from rest through a potential difference of 120 V. What is the de Broglie
 wavelength associated with the electron?

 (a) 0.014 nm   (b) 0.11 nm   (c) 1.1 mm   (d) 1.4 nm   (e) 8.6 nm

4. What is the maximum number of electrons that can occupy the d subshell?

 (a) 2   (b) 6   (c) 10   (d) 14   (e) 18

5. The probability of finding an electron at a given location in a hydrogen atom is directly proportional to
(a) its energy. (b) its momentum. (c) its angular momentum.
(d) the wave function. (e) the square of the wave function.

6. The element krypton has 36 electrons. How many electrons are in the $n = 5$ shell?
(a) 2 (b) 4 (c) 8 (d) 10 (e) 18

7. If the principal quantum number is $n = 4$, which of the following is not an allowed orbital quantum number?
(a) 4 (b) 3 (c) 2 (d) 1 (e) 0

8. A positron is a particle that has
(a) the same mass as an electron. (b) the same mass as a proton. (c) the same mass as a neutron.
(d) negative electronic charge. (e) no electronic charge.

9. A proton has kinetic energy of 1.0 MeV. If its momentum is measured with an uncertainty of ±1.0%, what is the uncertainty in its position?
(a) 2.8×10^{-14} m (b) 5.6×10^{-14} m (c) 9.1×10^{-14} m (d) 2.3×10^{-13} m (e) 9.1×10^{-13} m

10. Which one of the following particles will tend to have the longest de Broglie wavelength (assuming they travel at the same speed)?
(a) car (b) basketball (c) neutron (d) proton (e) electron

11. With what accuracy would you have to measure the position of a moving electron so its velocity was uncertain by 0.500 cm/s?
(a) 2.32 cm (b) 4.64 cm (c) 14.6 cm (d) 2.32 m (e) 4.64 m

12. What is the de Broglie wavelength for the matter wave associated with a 1200-kg car traveling at 90 km/h?
(a) 6.1×10^{-39} m (b) 2.2×10^{-38} m (c) $5/5 \times 10^{-37}$ m (d) 3.0×10^{4} m (e) 1.1×10^{5} m

Answers to Practice Quiz:

1.a 2.c 3.b 4.c 5.e 6.b 7.a 8.a 9.d 10.e 11.a 12.b

CHAPTER 29

The Nucleus

I.   Chapter Objectives

Upon completion of this chapter, you should be able to:

1.   distinguish between the Thompson and Rutherford-Bohr models of the atom, specify some of the basic properties of the strong nuclear force, and understand nuclear notation.

2.   define the term *radioactivity*, distinguish among alpha, beta, and gamma decay, and write nuclear-decay equations.

3.   explain the concepts of activity, decay constant, and half-life of a radioactive sample, and use radioactive decay to find the age of objects.

4.   state which proton- and neutron-number combinations result in stable nuclei, explain the pairing effect and magic numbers in relation to nuclear stability, and calculate nuclear binding energy.

5.   gain insight into the operating principles of various nuclear-radiation detectors, investigate the medical and biological effects of radiation exposure, and study some of the practical uses and applications of radiation.

II.   Chapter Summary and Discussion

1.   Nuclear Structure and the Nuclear Force (Section 29.1)

The **Rutherford-Bohr model** of the atom is essentially a planetary model, with negative electrons orbiting the positively charged nucleus in which almost all the mass of the atom is located. The atomic nucleus is composed of two types of particles—protons and neutrons—which are collectively referred to as **nucleons**. From scattering experiments, the *upper limit* for the nuclear radius was found to be on the order of 10^{-12} m.

The **strong nuclear force**, or simply the nuclear force, is the short-range attractive force between nucleons that is responsible for holding the nucleus together against the repulsive electric force of its protons. This force possesses the following properties:

- It is strongly attractive and much larger in relative magnitude than the electrostatic and magnetic forces.

- It is very short range; that is, a nucleon interacts only with its nearest neighbors, over a distance on the order of 10^{-15} m.

- It acts between any two nucleons within a short range, that is, between two protons, a proton and a neutron, or two neutrons.

In nuclear notation $^A_Z X_N$, Z is the **atomic number** or **proton number** (the number of protons), N is the **neutron number** (the number of neutrons), A is the **mass number** (the number of nucleons), and X is the chemical symbol of the element of the nucleus. The neutron number is often omitted because it can be determined from $N = A - Z$. For example, the $^{13}_6 C$ represents carbon-13, which has 6 protons, $13 - 6 = 7$ neutrons, and 13 nucleons.

Isotopes are nuclei with the same number of protons (atomic number) but a different number of neutrons (neutron numbers or mass numbers), such as $^{12}_6 C$, $^{13}_6 C$, and $^{14}_6 C$. A particular nuclear species or isotope of any element is called a **nuclide**, for example, $^{11}_6 C$, $^{12}_6 C$, $^{13}_6 C$, $^{14}_6 C$, $^{15}_6 C$, and $^{16}_6 C$ are all nuclides of carbon.

Example 29.1   The nuclear notation for cobalt-59 is $^{59}_{27} Co$. If the atom is neutral,

(a) how many protons does it possess?

(b) how many neutrons does it possess?

(c) how many nucleons does it possess?

(d) how many electrons does it possess?

Solution: Given: $A = 59$, $Z = 27$.

Find: Number of (a) protons,   (b) neutrons,   (c) nucleons,   and   (d) electrons.

From $A = N + Z$, $N = A - Z = 59 - 27 = 32$.

(a) The number of protons is equal to the atomic number, $Z = 27$ protons.

(b) The number of neutrons is equal to the neutron number, $N = A - Z = 59 - 27 = 32$ neutrons.

(c) The number of nucleons is equal to the mass number, $A = 59$ nucleons.

(d) In a neutral atom, the number of electrons is equal to the number of protons, so there are 27 electrons.

Example 29.2   Which one of the following is an isotope of the hypothetical nuclide $^{39}_{20} X_{19}$?

(a) $^{40}_{20} X_{20}$ (b) $^{39}_{19} Y_{20}$ (c) $^{39}_{21} Z_{18}$.

Solution:

Isotopes are nuclei with the same number of protons (Z) and a different number of neutrons (N). So the answer is (a). That is, $^{40}_{20} X_{20}$ and $^{39}_{20} X_{19}$ are isotopes.

---

2. Radioactivity (Section 29.2)

The nuclei of some isotopes are not stable, and they decay spontaneously with the emission of particles. Such isotopes are said to be *radioactive* or to exhibit **radioactivity**. Five commonly occurring decay modes are:

- **Alpha decay,** in which the nucleus emits an **alpha particle**, a doubly charged ($+2e$) particle containing two protons and two neutrons (helium nucleus, $_2^4\text{He}$).

- **Beta decay** (β^-) in which the nucleus emits a **beta particle**, or an electron ($_{-1}^0\text{e}$).

- **Gamma decay,** in which an excited nucleus emits a **gamma ray**, or a "particle" or quantum, or a photon, of electromagnetic energy.

- **Beta decay** (β^+) **decay**, in which the nucleus emits a positron ($_{+1}^0\text{e}$).

- **Electron capture**, in which the nucleus *absorbs* one of its atomic orbital electrons.

Alpha decay involves the quantum-mechanical process of **tunneling** or **barrier penetration**; that is, there is a finite probability of finding the alpha particle outside the nucleus after it penetrates the Coulomb potential energy barrier. In basic beta β^- decay, a neutron decays to a proton and an electron, $_0^1\text{n} \rightarrow _1^1\text{p} + _{-1}^0\text{e}$; and in basic beta β^+ decay, a proton decays to a neutron and a positron, $_1^1\text{p} \rightarrow _0^1\text{n} + _{+1}^0\text{e}$. (Neutrinos are also involved in beta decays. They will be discussed later and are not shown here for simplicity.)

Two laws apply to all nuclear reactions: the **conservation of nucleons** (the total number of nucleons remains constant) and the **conservation of charge** (the total charge remains constant).

Example 29.3   Write the nuclear equation for the following nuclear decays:

       (a) alpha decay of $_{84}^{214}\text{Po}$

       (b) beta (β^-) decay of $_6^{14}\text{C}$

       (c) gamma decay of $_{28}^{61}\text{Ni}^*$ (the * represents an excited nucleus)

Solution:

For every nuclear reaction, the total number of nucleons and the total charge remain constant.

(a) If $_{84}^{214}\text{Po}$ emits an alpha particle ($_2^4\text{He}$), the daughter nucleus must have $214 - 4 = 210$ nucleons (conservation of nucleons) and $84 - 2 = 82$ protons (conservation of charge). From the periodic table, the nucleus that has 82 protons is lead (Pb). Therefore, the nuclear decay equation is $_{84}^{214}\text{Po} \rightarrow _{82}^{210}\text{Pb} + _2^4\text{He}$.

(b) If $^{14}_6C$ emits a beta particle or electron ($^0_{-1}e$), the daughter nucleus must have the same number of nucleons (14) and $6 + 1 = 7$ protons to conserve nucleons and charge. The element with 7 protons is nitrogen (N). Therefore, the nuclear equation is $^{14}_6C \rightarrow ^{14}_7N + ^0_{-1}e$.

(c) Because a gamma ray (γ) is a photon, the daughter nucleus is still Ni, but at a lower energy level. The nuclear equation is $^{61}_{28}Ni^* \rightarrow ^{61}_{28}Ni + \gamma$.

The absorption or degree of penetration of nuclear radiation is an important consideration in applications such as radioisotope treatment of cancer and nuclear shielding around a nuclear reactor or radioactive materials. Alpha particles have large mass and are doubly charged and generally move slowly. A few centimeters of air or a sheet of paper will usually completely stop them. Beta particles can travel a few meters in air or a few centimeters in aluminum before being stopped. Gamma rays are more penetrating. They can penetrate a centimeter or more of a dense material such as lead.

3. Decay Rate and Half-Life (Section 29.3)

The **activity** of a radioactive isotope is defined as the number of disintegrations or decays (ΔN) per unit time (Δt), $\Delta N / \Delta t$. This activity is proportional to the number of undecayed nuclei present (N); that is, $\Delta N / \Delta t = -\lambda N$, where λ is the **decay constant**, which is different for different isotopes. The negative sign indicates the decrease in the number of undecayed nuclei present in a decay process. Solving the activity equation yields the expression for the undecayed nuclei as a function of time, $N = N_0 e^{-\lambda t}$, where N_0 is the initial number of nuclei at $t = 0$.

The **half-life** ($t_{1/2}$) of an isotope is the time required for the number of undecayed nuclei in a sample (hence the radioactive decay rate or activity of the sample) to fall to half of its original value. It is related to the decay constant λ through $t_{1/2} = 0.693 / \lambda$. The SI unit of activity is the **becqerel** (Bq), and 1 Bq = 1 decay/s. Another common unit is the **curie** (Ci), and 1 Ci = 3.70×10^{10} decays/s = 3.70×10^{10} Bq.

Note: To use the equation $N = N_0 e^{-\lambda t}$ it is not necessary to change time to seconds. You can use any unit (min, h, or year) as long as λ is in the same (inverse) units; for example, ($h^{-1} \times h$) = unitless.

Radioactive isotopes can be used as nuclear clocks. A common radioactive dating method used on materials that were once part of living things is **carbon-14 dating**. This method uses the fact that the carbon-14 content decreases with time once a living organism dies, so measurements of the concentration of carbon-14 in dead matter relative to that in living things can then be used to establish when the organism died.

Integrated Example 29.4

A radioactive isotope sample has an half-life of 10 min. (a) After 20 minutes, the fraction of the decayed nuclei is (1) zero, (2) 1/2, (3) 1/4, or (4) 3/4. Explain. (b) Use the decay constant to determine the fractions of the decayed nuclei after 20 and 55 min.

(a) Conceptual Reasoning:

20 min is equal to two half-lives. After one half-life, the fraction of nuclei remaining is $\frac{N_1}{N_o} = \frac{1}{2} = \frac{1}{2^1}$;

after two half-lives, the fraction of nuclei remaining is $\frac{N_2}{N_o} = \frac{N_2}{N_1} \times \frac{N_1}{N_o} = \frac{1}{2} \times \frac{1}{2} = \frac{1}{2^2} = \frac{1}{4}$. So the fraction of

decayed nuclei is $1 - = \frac{1}{4} = \frac{3}{4}$.

(b) Quantitative Reasoning and Solution:

Given: $t_{1/2} = 10$ min, $t = 20$ min and 55 min. Find: $1 - \frac{N}{N_o}$.

The decay constant is $\lambda = \frac{0.693}{t_{1/2}} = \frac{0.693}{10 \text{ min}} = 0.0693 \text{ min}^{-1}$.

For $t = 20$ min, $\lambda t = (0.0693 \text{ min}^{-1})(20 \text{ min}) = 1.386$.

From $N = N_o e^{-\lambda t}$, we have $\frac{N}{N_o} = e^{-\lambda t} = e^{-1.386} = 0.25 = \frac{1}{4}$.

So the fraction of decayed nuclei is $1 - \frac{1}{4} = \frac{3}{4}$, as expected.

For $t = 55$ min, $\lambda t = (0.0693 \text{ min}^{-1})(55 \text{ min}) = 3.812$.

$\frac{N}{N_o} = e^{-3.812} = 0.022$ so the fraction of decayed nuclei is $1 - 0.022 = 0.978$.

The answer for $t = 55$ min can also be obtained using half-life directly. After n half-lives, the fraction of

nuclei remaining is $\frac{N}{N_o} = \frac{1}{2^n}$. With $n = \frac{55 \text{ min}}{10 \text{ min}} = 5.5$ half-lives, $\frac{N}{N_o} = \frac{1}{2^{5.5}} = 0.022$. So the fraction of decayed

nuclei is $1 - 0.022 = 0.978$.

Example 29.5 A radioactive sample with a half-life of 10.0 min initially is composed of 1.50×10^{10} nuclei.

(a) What is the decay constant?

(b) What is the initial activity in Ci?

(c) What is the activity in Ci after 2.00 min?

Solution: Given: $N_0 = 1.50 \times 10^{10}$ nuclei, $t_{1/2} = 10.0$ min.

Find: (a) λ (b) $\left|\dfrac{\Delta N}{\Delta t}\right|_0$ (c) $\left|\dfrac{\Delta N}{\Delta t}\right|$.

(a) $\lambda = \dfrac{0.693}{t_{1/2}} = \dfrac{0.693}{10.0 \text{ min}} = 0.0693 \text{ min}^{-1} = 1.16 \times 10^{-3} \text{ s}^{-1}$.

(b) 1 Ci $= 3.70 \times 10^{10}$ decays/s.

$\left|\dfrac{\Delta N}{\Delta t}\right|_0 = \lambda N_0 = (1.16 \times 10^{-3} \text{ s}^{-1})(1.50 \times 10^{10}) = 1.74 \times 10^7 \text{ decays/s} = 4.70 \times 10^{-4} \text{ Ci}$.

(c) $\lambda t = (0.0693 \text{ min}^{-1})(2.00 \text{ min}) = 0.139$, so $N = N_0 e^{-\lambda t} = (1.50 \times 10^{10}) e^{-0.139} = 1.31 \times 10^{10}$ nuclei.

Therefore, $\left|\dfrac{\Delta N}{\Delta t}\right| = \lambda N = (1.16 \times 10^{-3} \text{ s}^{-1})(1.31 \times 10^{10}) = 1.52 \times 10^7 \text{ decays/s} = 4.11 \times 10^{-4} \text{ Ci}$.

---

4.   Nuclear Stability and Binding Energy (Section 29.4)

Experiments show that to lower their total energy in a nucleus, two protons will "pair up," as will two neutrons. This so-called **pairing effect** influences the following criteria for nuclear stability:

(1)   All isotopes with 83 protons or more ($Z \geq 83$) are unstable.

(2)   Most even-even (even number of protons and even number of neutrons) are stable. Many odd-even or even-odd nuclei are stable. Only four odd-odd nuclei are stable, ^{2}H, ^{6}Li, ^{10}Be, and ^{14}N.

(3)   Stable nuclei with fewer than 40 nucleons ($A < 40$) have approximately the same number of protons and neutrons. Stable nuclei with $A > 40$ have more neutrons than protons.

Example 29.6   Is the sodium isotope $^{22}_{11}$Na likely to be stable?

Solution:

(1)   criterion (1) is *satisfied* with $Z = 11$.

(2)   criterion (2) is *not satisfied*. The isotope $^{22}_{11}$Na has an odd-odd nucleus, and since it is not one of the four stable odd-odd nuclei, it must be unstable.

Because the masses of nuclei are so small, another unit, the **atomic mass unit (u)** is used to measure nuclear masses. By definition, 1 u $= 1.66054 \times 10^{-27}$ kg, and it is referenced to a neutral atom of ^{12}C, which is taken to have an exact mass of 12.000000 u. According to mass-energy equivalence, 1 u is equivalent to 931.5 MeV; that is 1 u of mass if converted completely would yield 931.5 MeV of energy.

It is found that the sum of the masses of the nucleons in a nucleus (bound particles) is less than that of the nucleons separated as free particles. This mass difference, or **mass defect** (Δm), has an energy equivalence called the total **binding energy** (E_b), which is the minimum amount of energy needed to separate a given nucleus into its constituent nucleons, $E_b = (\Delta m)c^2$.

The *average binding energy per nucleon*, E_b/A, gives an indication of nuclear stability. The greater the E_b/A, the more tightly bound are the nucleons in a nucleus, hence, the more stable it is.

Example 29.7   What is the total binding energy and average binding energy per nucleon for sodium $^{22}_{11}$Na. The mass of $^{22}_{11}$Na is 21.994435 u.

Solution: Sodium $^{22}_{11}$Na has 11 protons and 11 neutrons.

Given: $^{22}_{11}$Na atomic mass = 21.994435 u, $^1_1$H mass = 1.007825 u, $^1_0$n mass = 1.008665 u.

Find: E_b and E_b/A.

The mass of the separated nucleons is $m = 11(1.007825\ \text{u}) + 11(1.008665\ \text{u}) = 22.18139\ \text{u}$.

So the mass defect is $\Delta m = 22.18139\ \text{u} - 21.994435\ \text{u} = 0.18696\ \text{u}$.

Therefore, the total binding energy is $E_b = (0.18696\ \text{u})(931.5\ \text{MeV/u}) = 174.15\ \text{MeV}$.

The average binding energy per nucleon is $\dfrac{E_b}{A} = \dfrac{174.15\ \text{MeV}}{22\ \text{nucleons}} = 7.916\ \text{MeV/nucleon}$.

---

5.   Radiation Detection and Applications (Section 29.5)

A *Geiger counter*, a *scintillation counter*, a *solid-state (semiconductor) detector*, a *cloud chamber*, a *bubble chamber*, and a *spark chamber* are all examples of **radiation detectors**.

An important consideration in radiation therapy and radiation safety is the amount, or *dose*, of radiation. Several quantities are used to describe this amount in terms of *exposure*, *absorbed dose*, and *equivalent dose*. The earliest unit of dosage, the **roentgen** (R) was based on exposure. The roentgen has largely been replaced by the rad (*r*adiation *a*bsorbed *do*se), which is an absorbed dose unit. The SI unit for absorbed dose is the **gray** (Gy): 1 Gy = 1 J/kg = 100 rad.

The effective dose is measured in terms of the **rem** unit (*r*oentgen or *r*ad *e*quivalent *m*an). The different degrees of effectiveness of different particles are characterized by the **relative biological effectiveness (RBE)**. Then, effective dose (in rem) = dose (in rad) × RBE. Another unit of effective dose is the **sievert** (Sv), and effective dose (in Sv) = dose (in Gy) × RBE. Because 1 Gy = 100 rad, it follows that 1 Sv = 100 rem.

Some other applications of radiation in domestic and industrial use are the smoke detector, radioactive tracer (to detect leaks), and in **neutron activation analysis** (to activate compounds for detection purposes).

III. Mathematical Summary

Number of Undecayed Nuclei	$N = N_0 e^{-\lambda t}$ (29.3)	Calculates the number of undecayed nuclei as a function of time.
Half-Life and Decay Constant	$t_{1/2} = \dfrac{0.693}{\lambda}$ (29.4)	Relates the half-life and the decay constant of a radioisotope.
Activity of a Radioisotope	$R = \text{activity} = \left\lvert \dfrac{\Delta N}{\Delta t} \right\rvert = \lambda N$ (29.2)	Defines the activity of a radioisotope.
Total Binding Energy	$E_b = (\Delta m)c^2$ (29.5)	Computes the binding energy of a nucleus from the mass defect between the nucleus and the constituent particles.
Effective Dose	Dose (in rem) = dose (in rad) × RBE (29.6) Dose (in Sv) = dose (in Gy) × RBE (29.7)	Calculates the effective dose by including the relative biological effectiveness (RBE).

IV. Solutions of Selected Exercises and Paired Exercises

8. The mass number is $8 + 8 = 16$, $8 + 9 = 17$, and $8 + 10 = 18$, respectively.

    They are $\boxed{{}^{16}_{8}\text{O}, {}^{17}_{8}\text{O}, {}^{18}_{8}\text{O}}$.

12. (a) Isotope of an element has the same $\boxed{\text{(1) atomic number}}$.

    (b) Isotopes have the same $Z = 79$ but different $A = 197$.

    For example, $\boxed{{}^{196}_{79}\text{Au}, {}^{198}_{79}\text{Au}}$.

20. Due to momentum conservation, the decay particles are moving in opposite directions so they carry away some of the kinetic energy.

22. (a) $\boxed{{}^{60}_{27}\text{Co} \rightarrow {}^{60}_{28}\text{Ni} + {}^{0}_{-1}\text{e}}$.

     (b) $\boxed{{}^{226}_{88}\text{Ra} \rightarrow {}^{222}_{86}\text{Rn} + {}^{4}_{2}\text{He}}$.

26. (a) $\boxed{{}^{8}_{4}\text{Be}}$.

     (b) $\boxed{{}^{97}_{38}\text{Sr}}$.

     (c) $\boxed{{}^{47}_{21}\text{Sc}}$.

     (d) $\boxed{{}^{0}_{-1}\text{e}}$.

34. No, decay is exponential, not linear.

38. (a) The answer is $\boxed{\text{(2) one-eighth}}$ would be left, because 3 h is 3 half-lives, then $\left(\frac{1}{2}\right)^3$ or $\frac{1}{8}$.

     (b) 1 d = 24 $t_{1/2}$, so $\boxed{\dfrac{1}{2^{24}}, \text{ or approximately } 6 \times 10^{-6}\,\% \text{ of the original}}$.

40. $\lambda = \dfrac{0.693}{t_{1/2}} = \dfrac{0.693}{18 \times 60 \text{ s}} = 6.42 \times 10^{-4} \text{ s}^{-1}$;

     $\lambda t = (5.92 \times 10^{-4} \text{ s}^{-1})(3600 \text{ s}) = 2.31$.

     Thus, $\dfrac{\Delta N}{\Delta t} = \lambda N = \lambda N_0\, e^{-\lambda t} = \dfrac{\Delta N_0}{\Delta t}\, e^{-\lambda t} = (10 \text{ mCi})\, e^{-2.31} = \boxed{0.99 \text{ mCi}}$.

46. $\dfrac{\Delta N}{\Delta t} = \lambda N = \lambda N_0\, e^{-\lambda t} = \dfrac{\Delta N_0}{\Delta t}\, e^{-\lambda t}$, so $\dfrac{\Delta N/\Delta t}{\Delta N_0/\Delta t} = 0.20 = e^{-\lambda t}$.

     Thus, $t = -\dfrac{\ln 0.20}{\lambda} = -\dfrac{t_{1/2} \ln 0.20}{0.693} = -\dfrac{(5.3 \text{ years}) \ln 0.20}{0.693} = \boxed{12 \text{ years}}$.

51. $\lambda t = \dfrac{0.693}{t_{1/2}}\, t = \dfrac{0.693}{1600 \text{ y}} (2100 \text{ y} - 1898 \text{ y}) = 0.0875$.

     From $N = N_0\, e^{-\lambda t}$, we have $\dfrac{N}{N_0} = e^{-\lambda t} = e^{-0.0875} = 0.916$.

     Thus, the amount of radium that would remain is $(0.916)(10 \text{ mg}) = \boxed{9.2 \text{ mg}}$.

54.　(a) The end product is $\boxed{(2)\ {}^{13}\text{C}}$, because the decay is ${}^{13}_{7}\text{N} \rightarrow {}^{13}_{6}\text{C} + {}^{0}_{+1}\text{e}$.

(b) There were $N_o = \dfrac{0.0015\text{ kg}}{(13\text{ u})(1.66 \times 10^{-27}\text{ kg/u})} = 6.95 \times 10^{22}$ nuclei.

$\lambda = \dfrac{0.693}{t_{1/2}} = \dfrac{0.693}{10\text{ min}} = 0.0693\text{ min}^{-1}$.　$\lambda t = (0.0693\text{ min}^{-1})(35\text{ min}) = 2.43$.

$\dfrac{\Delta N_o}{\Delta t} = \lambda N_o = (0.0693\text{ min}^{-1})(6.95 \times 10^{22}\text{ nuclei}) = 4.82 \times 10^{21}$ decays/min.

$\dfrac{\Delta N}{\Delta t} = \dfrac{\Delta N_o}{\Delta t}\,e^{-\lambda t} = (4.82 \times 10^{21}\text{ decays/min})\,e^{-2.43} = \boxed{4.2 \times 10^{20}\text{ decays/min}}$.

(c) $N = \dfrac{\Delta N/\Delta t}{\lambda} = \dfrac{4.23 \times 10^{20}\text{ decays/min}}{0.0693\text{ min}^{-1}} = 6.10 \times 10^{21}$ nuclei.

So the percentage of ${}^{13}\text{N}$ is $\dfrac{6.10 \times 10^{21}}{6.95 \times 10^{22}} = \boxed{8.8\%}$.

63.　$E_b = \Delta m c^2 = (m_n + m_p - m_D)c^2$, so

$m_D = m_p + m_n - E_b/c^2 = 1.007276\text{ u} + 1.008665\text{ u} + \dfrac{2.224\text{ MeV}}{931.5\text{ MeV/u}} = \boxed{2.013\ 553\text{ u}}$.

66.　$E_b = \Delta m c^2 = (8m_H + 8m_n - m_O)c^2 = [8(1.007825\text{ u}) + 8(1.008665\text{ u}) - 15.994915\text{ u}](931.5\text{ MeV/u})$

$\qquad = 127.6$ MeV.

$\dfrac{E_b}{A} = \dfrac{127.6\text{ MeV}}{16\text{ nucleon}} = \boxed{7.98\text{ MeV/nucleon}}$.

72.　For Al: $E_b = (13m_H + 14m_n - m_{Al})c^2 = [13(1.007825\text{ u}) + 14(1.008665\text{ u}) - 26.981541\text{ u}](931.5\text{ MeV/u})$

$\qquad\qquad = 225.0$ MeV.

Thus, $\dfrac{E_b}{A} = \dfrac{225.0\text{ MeV}}{27\text{ nucleon}} = \boxed{8.33\text{ MeV/nucleon}}$.

For Na: $E_b = (11m_H + 12m_n - m_{Na})c^2 = [11(1.007825\text{ u}) + 12(1.008665\text{ u}) - 22.989770\text{ u}](931.5\text{ MeV/u})$

$\qquad\qquad = 186.6$ MeV.

Thus, $\dfrac{E_b}{A} = \dfrac{186.6\text{ MeV}}{23\text{ nucleon}} = \boxed{8.11\text{ MeV/nucleon}}$.

Therefore, the nucleons are more tightly bound in $\boxed{\text{Al}}$ on average.

82.　(a) $E = m \times \text{Dose} = (0.200\text{ kg})(1.25\text{ rad}) = \boxed{0.250\text{ J}}$.

(b) Dose (in rem) = Dose (in rad) $\times$ RBE $= (1.25\text{ rad})(4) = 5.00$ rem, which exceeds the maximum

permissible radiation dosage with background radiation included. So the answer is $\boxed{\text{yes}}$.

86. (a) $\boxed{{}^{241}_{95}\text{Am} \rightarrow {}^{237}_{93}\text{Np} + {}^{4}_{2}\text{He}}$.

(b) $\lambda = \dfrac{0.693}{t_{1/2}} = \dfrac{0.693}{(432 \text{ y})(3.16 \times 10^7 \text{ s/y})} = 5.076 \times 10^{-11} \text{ s}^{-1}$.

The number of original nuclei is $N_\text{o} = \dfrac{10^{-7} \text{ kg}}{(241)(1.66 \times 10^{-27} \text{ kg})} = 2.50 \times 10^{17}$ nuclei.

The activity is $\Delta N/\Delta t = \lambda N_\text{o} = (5.076 \times 10^{-11} \text{ s}^{-1})(2.50 \times 10^{17} \text{ nuclei}) = \boxed{1.27 \times 10^7 \text{ decays/s}}$.

(c) $\lambda t = \dfrac{0.693}{432 \text{ y}} \times (20 \text{ y}) = 0.03208$. $N = N_\text{o}\, e^{-\lambda t} = (2.50 \times 10^{17} \text{ nuclei})\, e^{-0.03208} = 2.42 \times 10^{17}$ nuclei.

$\Delta N/\Delta t = \lambda N = (5.076 \times 10^{-11} \text{ s}^{-1})(2.42 \times 10^{17} \text{ nuclei}) = \boxed{1.23 \times 10^7 \text{ decays/s}}$

88. (a) $\boxed{(1)\ {}^{12}\text{C}}$ requires more energy because all neutrons have paired up. To remove a neutron, a pair has to be broken.

(b) To remove a neutron from ${}^{12}\text{C}$, the resulting nucleus is ${}^{11}\text{C}$ and a neutron.

$E = [11.011\,433 \text{ u} + 1.008\,665 \text{ u} - 12.000\,000 \text{ u}](931.5 \text{ MeV/u}) = \boxed{18.72 \text{ MeV}}$.

To remove a neutron from ${}^{13}\text{C}$, the resulting nucleus is ${}^{12}\text{C}$ and a neutron.

$E = [12.000\,000 \text{ u} + 1.00\,8665 \text{ u} - 13.003\,355 \text{ u}](931.5 \text{ MeV/u}) = \boxed{4.95 \text{ MeV}}$.

(c) $\lambda = \dfrac{1.24 \times 10^3 \text{ nm·eV}}{E \text{ (in eV)}}$. $\lambda_{12} = \dfrac{1.24 \times 10^3 \text{ nm·eV}}{18.72 \times 10^6 \text{ eV}} = \boxed{6.62 \times 10^{-14} \text{ m}}$.

$\lambda_{13} = \dfrac{1.24 \times 10^3 \text{ nm·eV}}{4.95 \times 10^6 \text{ eV}} = \boxed{2.51 \times 10^{-13} \text{ m}}$.

V. Practice Quiz

1. How many nucleons are there in ${}^{15}_{8}\text{O}$?

(a) 7   (b) 8   (c) 15   (d) 22   (e) 23

2. Which particle has the least mass?

(a) alpha   (b) beta (β^-) (c) beta (β^+) (d) nucleon   (e) gamma

3. A radioactive sample has a half-life of 4.0 min. What fraction of the sample is left after 20 min?

(a) 1/32   (b) 1/25   (c) 1/16   (d) 1/5   (e) zero

4. An atom has 98 protons and 249 nucleons. If it undergoes beta (β^-) decay, what are the numbers of protons and nucleons, respectively, in the daughter nucleus?

(a) 96, 245   (b) 98, 250   (c) 96, 245   (d) 99, 249   (e) 100, 249

5.   The half-life of radioactive iodine-137 is 8.0 days. How many iodine nuclei are necessary to produce an activity of 1.0 μCi?

(a) 2.9×10^9 (b) 4.6×10^9 (c) 3.7×10^{10} (d) 7.6×10^{12} (e) 8.1×10^{13}

6.   What is the average binding energy per nucleon for $^{197}_{79}$Au? The mass of $^{197}_{79}$Au is 196.96656 u.

(a) 6.8 MeV   (b) 7.3 MeV   (c) 7.7 MeV   (d) 7.9 MeV   (e) 8.3 MeV

7.   The nuclear equation $^{227}_{89}$Ac $\rightarrow$ $^{223}_{87}$Fr $+ {}^4_2$He, is for what type of decay?

(a) alpha   (b) beta (β^-)   (c) gamma   (d) electron capture   (e) beta (β^+)

8.   An X-ray technician takes an average of ten X rays per day and receives 2.5×10^{-3} rad per X ray. What is the total dose in rem the technician receives in 250 working days?

(a) 2.50 rem   (b) 5.00 rem   (c) 6.25 rem   (d) 7.75 rem   (e) 9.00 rem

9.   Which one of the following nuclei is likely to be unstable?

(a) $^4_2$He   (b) $^{27}_{13}$Al   (c) $^{13}_6$C   (d) $^{24}_{11}$Na   (e) $^2_1$H

10.   The binding energy of a nucleus is directly related to

(a) radioactivity.   (b) mass defect.   (c) too many neutrons.

(d) conservation of nucleons.   (e) alpha decay.

11.   The strong nuclear force

(a) binds the orbital electrons to the atomic nucleus.

(b) has a longer range than the gravitational force.

(c) acts only between identical particles.

(d) overcomes the repulsive force between protons in the nucleus.

12.   When a $^6_3$Li nucleus is struck by a proton, an alpha particle and another nucleus are released. What is this other nucleus?

(a) $^2_2$He   (b) $^3_2$He   (c) $^4_2$He   (d) $^5_2$He   (e) $^6_2$He

Answers to Practice Quiz:

1.c 2.e 3.a 4.d 5.c 6.d 7.a 8.c 9.d 10.b 11.d 12.b

CHAPTER 30

Nuclear Reactions and Elementary Particles

I. Chapter Objectives

Upon completion of this chapter, you should be able to:

1. use charge and nucleon conservation to write nuclear reaction equations, and understand and use the concepts of Q value and threshold energy to analyze nuclear reactions.

2. understand the process of nuclear fission, the nature and the cause of a nuclear chain reaction, and the basic principles involved in the operation of nuclear reactors.

3. explain the fundamental difference between fusion and fission, calculate energy releases in fusion reactions, and understand how fusion might eventually provide a source of electric energy.

4. explain why the neutrino is necessary to account for observed beta-decay data, specify some of the physical properties of neutrinos, and write complete beta-decay equations.

5. understand the quantum-mechanical description of forces and classify the various forces according to their strengths, properties, ranges, and virtual particles.

6. classify the elementary particles into families, and understand the different properties of the various families of elementary particles.

7. become familiar with the quark model and properties of quark, and understand how the quark model accounts for the properties of baryons and mesons.

8. become familiar with current attempts to unify the four fundamental forces, and understand why elementary-particle interactions might hold the key to understanding the very early evolution of the universe.

II. Chapter Summary and Discussion

1. Nuclear Reactions (Section 30.1)

In **nuclear reactions**, a particular nuclide is converted into a different one, which in general is a nuclide of a completely different element. Some reactions can be produced by energetic particles from radioactive sources or from **particle accelerators**, in which particles are accelerated to high speeds. The general form of a nuclear reaction equation is $\mathbf{A} + \mathbf{a} \rightarrow \mathbf{B} + \mathbf{b}$, which is written as $\mathbf{A(a, b)B}$, where uppercase letters represent the nuclei, and the lowercase letters represent the particles. $\mathbf{A}$ and $\mathbf{a}$ are the reactants and $\mathbf{B}$ and $\mathbf{b}$ are the products.

The **Q value** of a reaction or decay represents the energy released ($Q > 0$) or absorbed ($Q < 0$) in the process; it appears as a change in the total mass of the system, $Q = (m_A + m_a - m_B - m_b)c^2 = (\Delta m)c^2$. When the Q value of a reaction is positive ($Q > 0$), the mass of the reactants is greater than the mass of the products, energy is released in the form of kinetic energy, and the reaction is said to be **exoergic**. When the Q value of a reaction is negative ($Q < 0$), the mass of the products is greater than the mass of the reactants, energy is absorbed, and the reaction is said to be **endoergic**. All naturally occurring radioactive decay processes are exoergic.

In an endoergic reaction, the minimum energy required for the reaction to happen is called the **threshold energy** and is given by $K_{min} = \left(1 + \dfrac{m_a}{m_A}\right)|Q|$. A measure of the probability that a particular reaction will occur is called the **cross section** of the reaction.

Note: The mass-energy equivalence of 1 u equivalent to 931.5 MeV is often used in calculating Q values and threshold energies.

Example 30.1   Is the following reaction exoergic or endoergic?

$$^7\text{Li} \quad + \quad ^1\text{p} \quad \rightarrow \quad ^4\text{He} \quad + \quad ^4\text{He}$$
$$(7.016005 \text{ u}) \quad (1.007825 \text{ u}) \qquad (4.002603 \text{ u}) \quad (4.002603 \text{ u})$$

Solution: Given: $m_A = 7.016005$ u, $m_a = 1.007825$ u, $m_B = 4.002603$ u, $m_b = 4.002603$ u.

Find: Q.

We need to first find the Q value and then use the sign of the Q value to determine whether the reaction is exoergic or endoergic.

$Q = (\Delta m)c^2 = (m_A + m_a - m_B - m_b)c^2$

$= (7.016005 \text{ u} + 1.007825 \text{ u} - 4.002603 \text{ u} - 4.002603 \text{ u})(931.5 \text{ MeV/u}) = +17.35 \text{ MeV}$.

Because the Q value is positive, the reaction is exoergic.

Example 30.2   Find the threshold energy for the following reaction:

$$^{13}\text{C} \quad + \quad ^1\text{p} \quad \rightarrow \quad ^{13}\text{N} \quad + \quad ^1\text{n}$$
$$(13.003355 \text{ u}) \quad (1.007825 \text{ u}) \qquad (13.005739 \text{ u}) \quad (1.008665 \text{ u})$$

Solution: Given: $m_A = 13.003355$ u, $m_a = 1.007825$ u, $m_B = 13.005739$ u, $m_b = 1.008665$ u.

Find: K_{min}.

The Q value of the reaction is

$$Q = (\Delta m)c^2 = (m_A + m_a - m_B - m_b)c^2$$

$$= (13.003355 \text{ u} + 1.007825 \text{ u} - 13.005739 \text{ u} - 1.008665 \text{ u})(931.5 \text{ MeV/u}) = -3.003 \text{ MeV}.$$

Thus, the threshold energy is $K_{min} = \left(1 + \dfrac{m_a}{m_A}\right)|Q| = \left(1 + \dfrac{1.007825 \text{ u}}{13.003355 \text{ u}}\right)(3.003 \text{ MeV}) = 3.236 \text{ MeV}.$

The threshold energy is more than 3.003 MeV (the energy absorbed by the reaction) because the products in the reaction must have kinetic energies to conserve linear momentum.

2.   Nuclear Fission (Section 30.2)

In a nuclear **fission reaction**, a heavy nucleus divides into two lighter nuclei whose total mass is less than that of the heavy nuclei, with the emission of two or more neutrons. The energy liberated in a fission reaction is about 1 MeV per nucleon in fission products. Because there are billions of nuclei in even a tiny sample of fissionable material, the amount of energy released in nuclear fission can be enormous. A sustained release of nuclear energy can be accomplished by a **chain reaction**, in which neutrons from one fission reaction initiate more fission reactions, and the process multiplies, the number of neutrons doubling or tripling with each generation. When this process occurs uniformly with time, there is a very fast exponential growth of released energy (or a bomb). To have a sustained chain reaction, there must be a minimum, or **critical mass**, of the fissionable material. Basically, enough mass is needed so the neutrons from the previous reaction do not escape without causing more fission reactions.

Example 30.3   Estimate the energy released in the reaction $^{235}_{92}\text{U} + ^{1}_{0}\text{n} \rightarrow ^{140}_{54}\text{Xe} + ^{94}_{38}\text{Sr} + 2(^{1}_{0}\text{n})$.

Solution:

There are $140 + 94 = 234$ nucleons in the fission products, so the energy release is approximately $(1 \text{ MeV/nucleon})(234 \text{ nucleons}) \approx 234 \text{ MeV}$. This is the energy released per initial ^{235}U nucleus decay.

Example 30.4   Find the energy released in the following neutron-initiated fission reaction:

$$^{1}\text{n} \quad + \quad ^{235}\text{U} \quad \rightarrow \quad ^{141}\text{Ba} \quad + \quad ^{92}\text{Kr} \quad + \quad 3(^{1}\text{n})$$
$$(1.008665 \text{ u}) \ (235.043925 \text{ u}) \quad\quad (140.91420 \text{ u}) \ (91.9252 \text{ u}) \ (1.008665 \text{ u})$$

Solution:

The mass difference between the reactants and the products is

$\Delta m = 1.008665$ u $+ 235.043925$ u $- 140.91420$ u $- 91.9252$ u $-3(1.008665$ u$) = 0.1872$ u.

Thus, the energy released is $(0.1872$ u$)(931.5$ MeV/u$) = 174.4$ MeV.

Currently, the only type of practical **nuclear reactor** is based on the fission chain reaction. There are four key elements to a reactor: fuel rods, coolant, control rods, and a moderator. **Fuel rods** are tubes packed with pellets of enriched uranium oxide (or other fissionable material) located in the reactor core. A **coolant** is usually ordinary water (H_2O) or heavy water (D_2O) flowing around the fuel rods to remove the energy released from the fission chain reaction. The chain reaction and energy output of a reactor are controlled by means of boron or cadmium **control rods**, which can be inserted into or withdrawn from the reactor core. (Boron and cadmium are very effective in absorbing neutrons.) The water flowing around the fuel rods not only acts as a coolant but also as a **moderator**, which slows down the speed of the neutrons from fission before they initiate more fission reactions. Other materials, such as graphite (carbon), may be used as moderators.

A **breeder reactor** produces more fissionable materials than it consumes. U-235 is consumed in a regular reactor, and fissionable Pu-239 is produced from neutron reactions with the non-fissionable isotope of uranium, U-238.

Should the coolant be lost or stop flowing through the core of a reactor, the reactor will overheat. This could result in a **LOCA** (loss-of-coolant accident) and the **meltdown** of the core. The huge amount of energy released can cause explosions and the release of radioactive materials to the environment. Therefore, a safety shutdown must be started when a loss of coolant is detected, and emergency cooling water must be provided to keep the radioactive fragments from overheating.

3.   Nuclear Fusion (Section 30.3)

In a nuclear **fusion reaction**, light nuclei fuse together to form a heavier nucleus with less total mass than the original nuclei, with the release of energy. At the centers of some stars, four hydrogen nuclei (^{1}H) fuse to form helium (^{4}He). The energy released per fusion is about 24.7 MeV. Due to the light mass of H and He, fusion releases more energy than fission on a per unit mass basis.

Fusion requires high temperatures that ionize the fusion material into a gas of positive ions and negative electrons called **plasma**. Due to the high temperature of plasma, plasma confinement is a major problem. The problem of plasma confinement is being approached in several different ways. Among these are magnetic confinement and inertial confinement. In **magnetic confinement**, magnetic fields are used to hold the plasma in a confined space, a so-called magnetic bottle. In **inertial confinement**, pulsed laser, electron, or ion beams would be used to implode hydrogen fuel pellets, producing compression and high temperatures.

Example 30.5   Find the energy released in the following fusion reaction:

$$^2\text{H} \quad + \quad ^2\text{H} \quad \rightarrow \quad ^3\text{He} \quad + \quad ^1\text{n}$$

$$(2.014102 \text{ u}) \ (2.014102 \text{ u}) \qquad (3.016029 \text{ u}) \ (1.008665 \text{ u})$$

Solution:

The mass difference between the reactants and the products is

$\Delta m = 2.014102 \text{ u} + 2.014102 \text{ u} - 3.016029 \text{ u} - 1.008665 \text{ u} = 3.510 \times 10^{-3} \text{ u}.$

Thus, the energy released is $(3.510 \times 10^{-3} \text{ u})(931.5 \text{ MeV/u}) = 3.270 \text{ MeV}.$

---

4.   Beta Decay and the Neutrino (Section 30.4)

In a beta decay, there is an apparent violation of conservation of energy, linear momentum, and angular momentum (spin). To account for this apparent problem, an additional particle called the **neutrino** (ν_e) was proposed. Neutrinos have zero rest mass (recent experiments show possible small mass) and interact with matter by the weak nuclear force (to be discussed later). Neutrinos (ν_e) are produced in β^+ decays, and antineutrinos ($\bar{\nu}_e$) are produced in β^- decays. The complete basic reactions are

$$\beta^- \text{ decay:} \qquad \text{n} \rightarrow \text{p} + \text{e}^- + \bar{\nu}_e \qquad\qquad \beta^+ \text{ decay:} \qquad \text{p} \rightarrow \text{n} + \text{e}^+ + \nu_e.$$

5.   Fundamental Forces and Exchange Particles (Section 30.5)

There are only four known **fundamental forces**: the *gravitational force*, the *electromagnetic force*, the *strong nuclear force*, and the *weak nuclear force*.

In quantum mechanics, forces are visualized as being transmitted by the exchange of particles. These particles are created within the time allowed by the uncertainty principle and are called **virtual particles**. The fundamental forces are considered to be carried by these virtual **exchange particles**. The greater the mass of an exchange particle, the more energy required to create it, and the shorter the range of the particle. Therefore, the range of the force produced by this particle is inversely related to its mass.

The exchange particle for the electromagnetic force is a (virtual) **photon**. Because a photon has zero mass, its range is infinite. The strong nuclear force is associated with the **meson**. The relationship between the mass of the meson and the range of the strong nuclear force is given by $R < c\Delta t = \dfrac{h}{2\pi m_m c}$, where m_m is the mass of the exchange particle. The mass of π mesons (**pions**) is found to be from 247 to 264 times that of an electron, so its range is about 10^{-15} m, which indicates a very short range force. The **W particle** carries the **weak nuclear force**. The W particle has a mass of about 100 times that of a proton, which explains the extremely short range ($\approx 10^{-17}$ m) and the weakness of the weak force. The weak force is the only force that acts on neutrinos, which explains why they are so difficult to detect. The **graviton** is the massless exchange particle for the gravitational force and interacts very weakly with matter, making it very difficult to detect. There is still no firm evidence of the existence of this massless particle.

A summary of the relative strength, range, and exchange particles of the four fundamental forces are given in the following table:

Force	Relative Strength	Action Distance	Exchange Particle
Strong Nuclear	1	Short range ($\approx 10^{-15}$ m)	Pion (π meson)
Electromagnetic	10^{-3}	Inverse square (infinite)	Photon
Weak Nuclear	10^{-18}	Extremely short range ($\approx 10^{-17}$ m)	W particle
Gravitational	10^{-45}	Inverse square (infinite)	Graviton

Integrated Example 30.6

The π^o meson has a mass 264 times that of an electron and the W particle has a mass 100 times that of an proton, respectively. (a) The range of the force mediated by the π^o meson is (1) greater than, (2) the same as, or (3) less than that of the W particle. Explain. (b) What are the ranges of the forces mediated by the π^o meson and W particle?

(a) Conceptual Reasoning:

Since the mass of the proton (1.67×10^{-27} kg) is about 1840 times that of the electron (9.11×10^{-31} kg), the mass of the π^o meson is much less than the mass of the W particle.

According to range of a force mediated by a particular particle, $R < c\Delta t = \dfrac{h}{2\pi m_m c}$, the range is inversely proportional to the mass of the particle. Therefore, the range of the force mediated by the π^o meson is (1) greater than that of the W particle.

(b) Quantitative Reasoning and Solution:

Given: m_m (meson) $= 264m_e = 264(9.11 \times 10^{-31}$ kg$) = 2.41 \times 10^{-28}$ kg,

m_m (W particle) $= 100m_p = 100(1.67 \times 10^{-27}$ kg$) = 1.67 \times 10^{-26}$ kg

Find: R for both meson and W particle.

Meson: $R < c\Delta t = \dfrac{h}{2\pi m_m c} = \dfrac{6.63 \times 10^{-34} \text{ J·s}}{2\pi(2.41 \times 10^{-28} \text{ kg})(3.00 \times 10^8 \text{ m/s})} = 1.46 \times 10^{-15}$ m.

W particle: $R < = \dfrac{6.63 \times 10^{-34} \text{ J·s}}{2\pi(1.67 \times 10^{-26} \text{ kg})(3.00 \times 10^8 \text{ m/s})} = 2.11 \times 10^{-17}$ m.

As expected, the range of the force mediated by the π^0 meson is (1) greater than that of the W particle.

---

6. Elementary Particles (Section 30.6)

The following **elementary particles** are fundamental particles, or the building blocks, of atoms: **leptons**, which are particles that interact by the weak nuclear force, and there are six of them, electrons, **muons**, **tauons**, and three types of neutrinos. Each lepton has its antiparticle, and so there are twelve different leptons in all.

Hadrons are particles that interact by the strong nuclear force, for example, protons, neutrons, and pions. The hadrons are subdivided into **baryons** and **mesons**. Baryons include the nucleons-protons and neutrons and have half-integer spin values ($\frac{1}{2}$ or $\frac{3}{2}$). Mesons, which include the pion, have integer spin values (0 or 1).

7. The Quark Model (Section 30.7)

and

8. Force Unification Theories, the Standard Model, and the Early Universe (Section 30.8)

Quarks are elementary particles that make up hadrons. Quarks combine only in two ways, either in threes or in quark-antiquark pairs. Three-quark combinations are called baryons, and quark-antiquark combinations are called mesons.

There are six *flavors* or types of quarks: up (u), down (d), strange (s), charm (c), top (t), and bottom (b). Quarks have fractional electronic charges of either $-e/3$ or $+2e/3$. Quarks interact by the strong nuclear force, but they are also subject to the weak force. A weak force acting on a quark changes its flavor and gives rise to the decay of hadrons.

The exchange particle for quarks is the **gluon**. To give the strong force a field representation, each quark is said to possess an analog of electric charge that is the source of the "gluon field." Instead of charge, the property of quarks is called **color charge** (no relationship with ordinary color). The force between quarks of different color is sometimes called **color force**. This theory is named *quantum chromodynamics* (QCD). Each quark can come in one of the three possible colors: red, blue, and green. There are corresponding anticolors and antiquarks. When a quark emits or absorbs a gluon, it changes its color.

The electromagnetic force and the weak force are two parts of a single **electroweak force**. Attempts to unify the various forces are called **unification theories**. A theory that would merge the strong nuclear force and the electroweak force is called the **grand unified theory** (GUT). Perhaps all forces are part of a single **superforce**, which is currently a primary theoretical challenge in elementary particle physics.

The **standard model** is the combination of the electroweak theory and quantum chromodynamics. In this model, the gluons carry the strong force, which keeps quarks together. Gravitons carry the gravitational interactions, and photons, W, and Z bosons carry the electroweak force.

It is currently believed that the universe began 10 to 15 billion years ago with a huge explosion named the *Big Bang*. The temperature during the first 10^{-45} second of the universe was on the order of 10^{32} K. This high temperature resulted in very high kinetic energy of the particles, so their rest masses were negligible. As the universe expanded and cooled (over billions of years the universe has cooled to the present 3 K) the particles condensed into what we see today. First, protons, neutrons, and electrons formed; in turn, they combined into atoms and, eventually, molecules.

III.   Mathematical Summary

Q Value (for reaction $A + a \rightarrow B + b$)	$Q = (m_A + m_a - m_B - m_b)c^2$ $= (\Delta m)c^2 \qquad (30.3)$	Defines the Q value of a nuclear reaction in terms of the mass defect.
Threshold Energy (stationary target only)	$K_{min} = \left(1 + \dfrac{m_a}{m_A}\right)|Q| \qquad (30.4)$	Calculates the minimum energy required (threshold energy) for an endoergic reaction.
Range of Exchange Particle	$R < c\Delta t = \dfrac{h}{2\pi m_m c} \qquad (30.5)$	Defines the range of an exchange particle.

IV. Solutions of Selected Exercises and Paired Exercises

8.   (a) $\boxed{^{37}_{16}\text{S}}$.   (b) $\boxed{^{135}_{52}\text{Te}}$.   (c) $\boxed{4(^1_0\text{n})}$.   (d) $\boxed{\text{p or } ^1_1\text{H}}$.   (e) $\boxed{^{137}_{56}\text{Ba}}$.

10.   (a) Initial mass $= m_C + m_H = 13.003\,355\text{ u} + 1.007\,825\text{ u} = 14.011\,180\text{ u}$.

Final mass $= m_{He} - m_B = 4.002\,603 + 10.012\,938\text{ u} = 14.015\,541\text{ u}$.

Since the final mass is greater, the reaction is $\boxed{\text{endoergic}}$, i.e., it requires energy input.

(b) $Q = (m_C + m_H - m_{He} - m_B)c^2 = (13.003\,355\text{ u} + 1.007\,825\text{ u} - 4.002\,603 - 10.012\,938\text{ u})(931.5\text{ MeV/u})$

$= -4.06\text{ MeV}$.

$K_{min} = \left(1 + \dfrac{m_a}{M_A}\right)|Q| = \left(1 + \dfrac{1.007\,825}{13.003\,355}\right)(4.06\text{ MeV}) = \boxed{4.32\text{ MeV}}$.

14. $Q = (m_O + m_n - m_C - m_{He})c^2 = (15.994\,915\text{ u} + 1.008\,665\text{ u} - 13.003\,355\text{ u} - 4.002\,603\text{ u})(931.5\text{ MeV/u})$

$= -2.215\text{ MeV}$.

$K_{min} = \left(1 + \dfrac{m_a}{M_A}\right)|Q| = \left(1 + \dfrac{1.008\,665}{15.994\,915}\right)(2.215\text{ MeV}) = \boxed{2.35\text{ MeV}}$.

21.   (a) Since $K_{min} = \left(1 + \dfrac{m_a}{M_A}\right)|Q|$, the difference in Q value (3 times) is greater than the difference caused

by the target mass (15 and 20). The first reaction has $\boxed{(1)\text{ greater}}$ minimum threshold energy.

(b) For the first reaction: $(K_1)_{min} = \left(1 + \dfrac{1}{15}\right)|Q_1|$,

For the second reaction: $(K_2)_{min} = \left(1 + \dfrac{1}{20}\right)|Q_2|$.

Thus, $\dfrac{(K_1)_{min}}{(K_2)_{min}} = \dfrac{16/15}{21/20}\dfrac{|Q_1|}{|Q_2|} = \dfrac{16/15}{21/20}(3) = \boxed{3.05}$.

28.   The energy liberated is about 1 MeV per nucleon in fission products.

(a) $235 + 1 - 5 = 231$ nucleons are involved in the fission process,

so the energy released is $(1\text{ MeV/nucleon})(231\text{ nucleon}) = \boxed{231\text{ MeV}}$.

(b) $235 + 1 - 3 = 233$ nucleons are involved in the fission process,

so the energy released is $(1\text{ MeV/nucleon})(233\text{ nucleon}) = \boxed{233\text{ MeV}}$.

30. (a) $Q = (2m_{H-2} - m_{H-3} - m_n)c^2 = [2(2.014\ 102\ u) - 3.016\ 029\ u - 1.008\ 665\ u](931.5\ MeV/u)$

$= \boxed{3.27\ MeV}$.

(b) $Q = (m_{H-2} + m_{H-3} - m_{He} - m_n)c^2$

$= (2.014\ 102\ u + 3.016\ 029\ u - 4.002\ 603\ u - 1.008\ 665\ u)(931.5\ MeV/u) = \boxed{17.6\ MeV}$.

37. Assuming the neutrino has no mass, $E = pc = \dfrac{hc}{\lambda}$.

Therefore $\lambda = \dfrac{hc}{E} = \dfrac{(6.63 \times 10^{-34}\ J\cdot)(3.00 \times 10^8\ m/s)}{(2.65 \times 10^6\ eV)(1.6 \times 10^{-19}\ J/eV)} = \boxed{4.69 \times 10^{-13}\ m}$.

38. (a) The beta particle is moving so it has kinetic energy. Also according to the conservation of energy, its

kinetic energy cannot exceed 5.35 MeV so the answer is $\boxed{(2)\ less\ than\ 5.35\ MeV\ but\ not\ zero}$.

(b) The total energy of the beta particle is $E = 5.35\ MeV - 2.65\ MeV = \boxed{2.70\ MeV}$.

The kinetic energy of the beta particle is $K = E - mc^2 = 2.70\ MeV - 0.511\ MeV = \boxed{2.19\ MeV}$.

The daughter nucleus has $\boxed{zero}$ kinetic energy.

(c) Since the daughter nucleus has zero kinetic energy in this case, the momentum of the beta particle is the

same as the neutrino's but in a direction that is $\boxed{exactly\ opposite\ that\ of\ the\ neutrino}$.

39. In a β^- decay: $\quad {}^A_Z P \rightarrow {}^A_{Z+1} D + {}^0_{-1} e$.

So $Q = (m_p - m_D - m_e)c^2 = \{(m_p + Zm_e) - [m_D + (Z+1)m_e]\}c^2 = (M_p - M_D)c^2$.

40. The daughter is ^{12}C. From Exercise 30.39,

$K_{max} = Q = (M_p - M_D)c^2 = (12.014\ 353\ u - 12.000\ 000\ u)(931.5\ MeV/u) = \boxed{13.37\ MeV}$.

44. (a) Use the results from Exercises 30.39 and 30.42.

For a reaction to be energetically possible, the Q value has to be positive.

In a β^- decay, $\quad Q = (M_p - M_D)c^2 > 0$, $\qquad$ so $\qquad M_p > M_D$.

In a β^+ decay, $\quad Q = (M_p - M_D - 2m_e)c^2 > 0$, $\qquad$ so $\qquad M_p > M_D + 2m_e$.

Therefore, M_p for β^- decay must be $\boxed{(2)\ less\ than}$.

(b) $\boxed{\beta^-: M_p > M_D;\ \beta^+: M_p > M_D + 2m_e}$.

48. Protons: strong, electromagnetic, gravitational.

    Electrons: electromagnetic, gravitational.

52. $m_m = 264m_e$, so $\Delta E = m_m c^2 = 264(0.511 \text{ MeV}) = \boxed{135 \text{ MeV}}$.

60. (a) The quark combination for a proton is $\boxed{(2) \; uud}$.

    (b) The proton has a charge of $+e$. $\frac{2}{3}e + \frac{2}{3}e - \frac{1}{3}e = +e$.

64. (a) $\boxed{{}^1_1 H}$. (b) $\boxed{{}^{59}_{28} Ni}$. (c) $\boxed{{}^{93}_{38} Sr}$.

    (d) $\boxed{n \text{ or } {}^1_0 n}$. (e) $\boxed{{}^{16}_8 O}$.

67. In an electron capture: $^A_Z P + {}^0_{-1} e \rightarrow {}^A_{Z-1} D$.

    So $Q = (m_p + m_e - m_D)c^2 = \{(m_p + Zm_e) - [m_D + (Z-1)m_e]\}c^2 = (M_p - M_D)c^2$.

68. (a) From Exercise 30.67, the Q value in electron capture is

    $Q = (m_{Be} - m_{Li})c^2 = (7.016930 \text{ u} - 7.016005 \text{ u})(931.5 \text{ MeV/u}) = 0.86 \text{ MeV}$.

    From Exercise 30.48, the Q value in β^+ decay is

    $Q = (M_p - M_d - 2m_e)c^2 = (7.016930 \text{ u} - 7.016005 \text{ u})(931.5 \text{ MeV/u}) - 2(0.511 \text{ MeV}) = -0.16 \text{ MeV}$.

    So the answer is $\boxed{(1) \text{ electron capture}}$ (β^+ decay is not possible).

    (b) If daughter recoil is ignored, the energy of the emitted neutrino is equal to the energy released in

    electron capture, i.e., $\boxed{0.86 \text{ MeV}}$.

V. Practice Quiz

1. What is the Q value of the following reaction?

   $$^{14}N \quad + \quad {}^4He \quad \rightarrow \quad {}^1p \quad + \quad {}^{17}O$$

   (14.003074 u) (4.002603 u) (1.007825 u) (16.999131 u)

   (a) 0 (b) -1.279×10^{-3} MeV (c) 1.279×10^3 MeV (d) -1.191 MeV (e) 1.191 MeV

2. Meson particles are made up of

   (a) three-quark combinations. (b) three-antiquark combinations.

   (c) quark-antiquark combinations. (d) two-quark combinations.

   (e) two-antiquark combinations.

3. The fuel for nuclear fission is

    (a) hydrogen.   (b) helium.   (c) uranium.   (d) neutron.   (e) any radioactive material.

4. Which one of the following is an example of a lepton?

    (a) electron   (b) proton   (c) neutron   (d) pion   (e) gluon

5. What is the energy released (positive) or absorbed (negative) in the following reaction:

    $$^3\text{H} \quad + \quad ^3\text{H} \quad \rightarrow \quad ^4\text{He} \quad + \quad 2(^1\text{n})$$

    (3.016049 u) (3.016049 u)   (4.002603 u)   (1.008665 u)

    (a) 0   (b) −0.0122 MeV   (c) 0.0122 MeV   (d) −11.3 MeV   (e) 11.3 MeV

6. Calculate the range of a hypothetical force if the electron were a virtual exchange particle.

    (a) 3.9×10^{-13} m   (b) 1.2×10^{-12} m   (c) 2.4×10^{-12} m   (d) 7.3×10^{-4} m   (e) 1.5×10^{-3} m

7. The charge of some quarks and antiquarks is

    (a) 0.   (b) $e/3$.   (c) $e/4$.   (d) $e/2$.   (e) e.

8. In a nuclear fussion, the mass of the reactants compared with the mass of the products is

    (a) zero.   (b) greater.   (c) smaller.   (d) the same.   (c) infinite.

9. A chain reaction can occur

    (a) in any uranium core.   (b) when critical mass is reached.   (c) in the center of the Sun.

    (d) when the control rods are fully inserted.   (e) when the coolant is too hot.

10. What is the threshold energy for the following reaction?

    $$^{14}\text{N} \quad + \quad ^4\text{He} \quad \rightarrow \quad ^1\text{p} \quad + \quad ^{17}\text{O}$$

    (14.003074 u) (4.002603 u) (1.007825 u) (16.999133 u)

    (a) 0   (b) −1.193 MeV   (c) 1.193 MeV   (d) −1.534 MeV   (e) 1.534 MeV

11. Which one of the following is *not* a property of a neutrino?

    (a) $m = 0$ (b) $v = 0$ (c) $v = c$ (d) $E = pc$ (e) spin $= \frac{1}{2}$

12. The magnetic force is part of the

    (a) electroweak force.   (b) week force.   (c) strong force.   (d) superforce.

Answers to Practice Quiz:

1. d 2. c 3. c 4. a 5. e 6. a 7. b 8. c 9. b 10. e 11. b 12. a